"The Atlanteans respected intelligence. They respected genius in any line of achievement, be it government, religion, philosophy, art, or science. As a result, the best and the brightest of the Atlantean geniuses had a much better chance than today to make a contribution." —Nikola Tesla

"The recuperative powers of the planet are enormous. If the pollution which has occurred is stopped, the damage that has been done will be recycled and the planet will recover. At this point, the damage is relatively insignificant."
—Luther Burbank

"Even something as commonplace as the popular music of today adds to the clutter of the astral. When you have millions of people feeding themselves on modern jazz and rock and roll, which are more violence and lust set to sound than they are genuine music, the astral problem that results influences not only the perceptions of psychics, but the feelings of ordinary people as well." —Sir Oliver Lodge

"We need to give the scientist the perspective that he is merely in the process of discovering what God has already thought of and created. True science is the discovery of divine vision—and the understanding of how this divine vision has manifested itself in the physical plane."
—Albert Einstein

OTHER BOOKS BY
ROBERT R. LEICHTMAN, M.D.

The Psychic Perspective
The Inner Side of Life
The Priests of God
The Dynamics of Creativity
The Destiny of America

FROM HEAVEN TO EARTH:

THE HIDDEN SIDE OF SCIENCE

BY ROBERT R. LEICHTMAN, M.D.

The Third In A Series

ARIEL PRESS
Atlanta, Georgia

This book is made possible by a gift
to the Publications Fund of Light
from Marilyn & Richard McCraney

THE HIDDEN SIDE OF SCIENCE
Copyright © 1979, 1980, & 1992 by Light

All Rights Reserved. No part of this book may be used or reproduced in any manner whatsoever without written permission, except in the case of brief quotations embodied in articles and reviews. Printed in the United States of America. Direct inquiries to: Ariel Press, 14230 Phillips Circle, Alpharetta, GA 30201. No royalties are paid on this book.

ISBN 0-89804-083-3

TABLE OF CONTENTS

A Brief Introduction — page 7
Nikola Tesla Returns — page 11
Luther Burbank Returns — page 73
Sir Oliver Lodge Returns — page 125
Einstein Returns — page 187
Glossary — page 261
From Heaven To Earth — page 271

A BRIEF INTRODUCTION

The Hidden Side of Science is the third of six books in a unique series by Dr. Robert Leichtman called *From Heaven to Earth*. Each book in this series contains the transcript of four interviews, conducted mediumistically, between Dr. Leichtman and the spirit of a well-known leader or genius.

The interviews began in 1973. They were first published, in 1979 and 1980, as a series of twelve books, each book containing one complete interview with the spirit of an outstanding psychic or medium—people such as Edgar Cayce, Arthur Ford, and Madame Blavatsky.

The popularity of the first series of interviews encouraged Dr. Leichtman to embark on a second set of twelve. This time, however, he decided to broaden the scope of the interviews so that it would embrace the whole domain of creativity, genius, and leadership. As a result, he interviewed such people as Albert Schweitzer, Mark Twain, and Albert Einstein. These interviews were conducted in 1979 and 1980 and published in 1981 and 1982.

Public acceptance of these interviews has remained strong for a decade. But it has proven awkward to keep 24 individual books in print, so we have decided to reissue all 24 interviews in a new format—six books each containing four interviews.

8 — From Heaven to Earth

The 24 interviews divided naturally into six groups, so that each book in the new format focuses on one specific theme of the work of bringing heaven to earth. The four interviews in this volume, for instance, all deal with the inner dimensions of science and the true nature of the scientific method. Nikola Tesla was the electrical genius who made practical the generation of alternating current, harnessed the enormous electrical potential of the Niagara Falls, and proved that electricity can be broadcast through the atmosphere just like radio waves. He talks about the need for science to think of the whole solar system as electrically charged in order to make the next great step forward in tapping new energy sources. Luther Burbank, the developer of such new plant varieties as the Burbank potato and the Shasta daisy, comments on the inner life of plants and their intimate interrelation with humanity. Sir Oliver Lodge, a university president and physicist whose research contributed to the development of the radio, also happened to be one of the leading explorers of mediumship in his day. He discusses the application of the scientific method to the study of psychic phenomena—and the need for science to drop its closed-minded prejudices against the psychic realities of life and begin studying them openly and honestly. Albert Einstein, of course, revolutionized scientific thinking with his theory of relativity. He presents an outline for the beginnings of a scientific cosmology—a scientific understanding of God and His creation.

The other five volumes in the new format are thematically grouped in much the same way:

The Psychic Perspective—interviews with Edgar Cayce, Eileen Garrett, Arthur Ford, and Stewart Edward White.

The Inner Side of Life—interviews with C.W. Leadbeater, H.P. Blavatsky, Cheiro, and Carl Jung and Sigmund Freud.

The Priests of God—interviews with Albert Schweitzer, Paramahansa Yogananda, the industrialist Andrew Carnegie, and Sir Winston Churchill.

The Dynamics of Creativity—interviews with William Shakespeare, Mark Twain, Rembrandt, and Richard Wagner.

The Destiny of America—interviews with Thomas Jefferson, Benjamin Franklin, Abraham Lincoln, and a group interview with seven key people from America's history—Alexander Hamilton, Jefferson, Franklin, the two Roosevelts, Harry Truman, and George Washington.

As the title *From Heaven to Earth* suggests, the purpose of these interviews is to acquaint readers with the current thinking of these outstanding individuals, even though they have left their physical bodies and now work on the inner dimensions of reality. Many new ideas about psychic phenomena, spiritual growth, government, art, service, and civilization are set forth in the conversations—as well as a new revelation of the relation between heaven and earth. The interviews are not just academic discussions of the historical accomplishments of these people; they probe new frontiers of the human mind. Each is a thoughtful, witty, and lively exchange of ideas.

It is not the intent of this series to document the existence of life after death—or the effectiveness of mediumship in contacting the spirits of those who have left their physical bodies. Nor is it necessary, for these matters have been scientifically proven many times over in other writings—indeed, in many of the books written by the people interviewed in these books. The doubting reader will find ample proof in the works of Sir Oliver Lodge, Stewart Edward White, Eileen Garrett, Madame Blavatsky, Arthur Ford, and C.W. Leadbeater—as well as the many books about Edgar Cayce. Instead, the interviews in *From Heaven to Earth* are offered as a way of demonstrating that we need not be content with just an echo of great geniuses who have lived and died; their voices can literally be heard again. Their spirits and ideas can actually return to earth. Heaven is not some faraway place inaccessible to mortals. It can easily be contacted by competent psychics and mediums who have correctly trained themselves. And such contact can produce insight and new ideas of great importance.

A more complete introduction to the interviews is contained in volume one, The Psychic Perspective, *in conjunction with the inter-*

view with Edgar Cayce. In it, the nature of the mediumistic trance, the origins of this project, and the value of creative genius are all discussed in detail. For information on ordering this book, or the entire series, please see page 271 in this volume.

—Carl Japikse
ARIEL PRESS

NIKOLA TESLA RETURNS

Born in Yugoslavia, Nikola Tesla landed in New York in 1884 at the age of 28, one of thousands of immigrants to come from Europe to the United States that year. He arrived with four pennies in his pocket—and his mind bursting with brilliant ideas which would stun the scientific community, introduce humanity to the true potential of energy, and light up the world.

In comparison to his contemporary, Thomas Edison, Nikola Tesla's name is not well known. Yet it should be, for without his scientific breakthroughs and inventions, our modern world would be very much different than it is. We would have Edison's light bulb, but only people living within a mile or two of a power station would be able to use it. For it was Tesla's development of alternating current generators, transformers, and motors that overcame the immense limitations of Edison's direct current systems and made the long-distance transmission of electricity practical. It was truly Tesla's genius which lit up the world—not just homes of the wealthy. Nor was Tesla's genius limited to just this one accomplishment, as staggering as it was. He invented the Tesla coil, which is still used widely today in radio and television sets and other electronic devices. He pioneered the use of remote control. He produced better turbines than anyone else. He created artificial lightning. He harnessed the electrical potential of Niagara Falls—a stupendous feat at the time.

In 1900, twenty years before the first commercial radio broadcast, he attempted to construct a mammoth tower on Long Island to broadcast not only radio and television waves, but also other electromagnetic energies. It was a grand scheme which would have revolutionized the shape of modern life even more than his other projects—but it fell through when financial support was withdrawn.

At the height of his success, Tesla maintained an elaborate laboratory in New York. From time to time, he opened the doors of the lab to the public and gave demonstrations about the nature of electricity and the new developments he was working on. The demonstrations were often quite spectacular, with Tesla calmly sending electric current through his body to light up an incandescent tube he was holding in his hand. The lab itself contained many innovations which were never commercially exploited. None of his lights or motors, for example, were connected to the power source by wires. Instead, a single loop ran around the four walls, near the ceiling. This loop was continuously electrified, and from it all of Tesla's lamps and motors somehow drew their power.

As impressive as Tesla's inventions and innovations, however, are the breakthroughs he made in the scientific understanding of the principles of electricity. More than anyone else before him or since, Tesla *comprehended* electricity. He knew how it behaved and what could be expected of it, not just from observing electrical phenomena, but also from directly perceiving, intuitively, the archetypal patterns of electricity.

It was this understanding of the nature of electricity which allowed Tesla to recognize the value of alternating current and how to make it practical—at a time when other scientists had pronounced it nothing more than a scientific curiosity, with no usable merit. It was likewise this profound understanding and insight which drove Tesla to conduct a series of experiments in Colorado Springs, where he proved the existence of terrestrial stationary waves—proved that the earth's atmosphere is electrically charged and can carry electromagnetic waves from any given point to any other point on the earth's surface. He considered this discovery the most important of his career, and put it to a practical test, broadcasting electricity through the air without wires and lighting lamps twenty-five miles away with it. Even scientists of today do not understand fully the significance of these particular discoveries.

Sadly, many of Tesla's best ideas never went beyond the experimental stage. A world occupied with its mundane concerns could not appreciate the treasures of Tesla's transcendental thinking while he was alive, and did not adequately support his work financially. Indeed, late in life, Tesla had hardly any funds to finance his scientific investigations. Yet he kept pursuing new understandings, even though the majority of his insights could not be applied.

One of his most spectacular ideas, though, did gain a certain amount of attention. He developed a "death ray" which was capable of sending destruction through the air up to distances of 250 miles. Recognizing the deadly potential of such a device, however, and the inability of modern governments to use it wisely, he kept the principles behind it a secret.

Tesla also experimented extensively with the principles of resonance, which is the capacity of the energies of one physical object to vibrate in step with the vibratory patterns of another object or force. He comprehended these principles well enough to be able to create an artificial earthquake with a tiny machine, and claimed he had the knowledge to split the earth like an apple. Today, many scientists are trying to rediscover, in their work, the principles of resonance which Tesla explored in his—but did not receive the support to develop.

In fact, a great mystique has been generated in certain circles about this man and his discoveries. Russian scientists reportedly are trying to duplicate his "death ray." It is darkly rumored that recent changes in weather patterns are the result of Soviet experiments with the principles pioneered by Tesla. American scientists and inventors, too, have taken the body of work left behind by Tesla at his death in 1943 and tried to explore its implications. Yet few are able to reproduce his thinking and understanding of scientific principles.

That is a point worth musing on. In the long run, it is not the inventions and the breakthroughs which should be considered Tesla's greatest contributions to mankind, as spectacular and as important as they were. Far more important was the style he set for scientific thinking and discovery—the way his mind worked. For Tesla truly was an example of enlightened genius and set a level of brilliance that should be the model to which all scientists, geniuses, and intelligent people strive.

To Tesla, thinking was a process which bridged the objective, physical world with several subtle realms, where ideas and concepts exist almost like fish in an ocean, waiting to be caught by the inquiring human mind. He was not satisfied with speculating about ideas and theories; he pushed beyond speculation and learned to deal with ideas and thoughts in their own realm, where they can be fully perceived.

In other words, Tesla used the mind to link heaven with earth. He is certainly not the first person ever to do this, but most other geniuses who do learn to bring heaven to earth do it in religious, philosophical, or creative ways. While their contributions are valuable, they are almost always couched in vague words and intangible feelings about love, light, and beauty. Tesla, by contrast, focused the ideas of the mind toward scientific discovery. He translated his ideas not into vague precepts and lovely thoughts, but into turbines, lighting systems, broadcasting towers, and electrical machinery. In his hands, the light of heaven literally became light on earth.

In the process, Tesla lifted the scientific method to new heights. It might be said he embodied the true spirit of science better than anyone else ever has. He set an example of what the enlightened scientist *ought to be*.

One of the key ingredients of his style of working, for example, was his marvelous curiosity. He sought to comprehend *all of life*. He did not impose pet theories or preconceived expectations upon his thinking, as so many modern scientists do. He carefully trained his mind to be objective and clearsighted, then let nature and the laws of physics speak for themselves. Frequently, his curiosity led him into intense periods of concentration, in which his thinking seemed to penetrate ideas and principles themselves, tapping a level of inspiration behind both the ideas and the observable phenomena associated with them.

Another important element in Tesla's approach to science was his fearless attitude about shattering scientific tradition. He was neither a radical nor a rebel, but he did not hesitate to press beyond the current limits of scientific understanding and perspective. He knew that life itself is not limited by science's restricted conceptions of it, and did not fear the smallmindedness of scientists who do seek to put limits on life. During his experiments in Colorado, for example, Tesla detected signals he thought came from other planets. Most scientists would be afraid to report such an observation, for fear of being ridiculed by their colleagues. Tesla, however, reported it. Predictably, the news was met with scorn in numerous "scientific" journals, thus cutting off further investigation.

Undoubtedly, such opposition from traditionalists must have upset Tesla to a degree. But unlike many innovators who get bogged down in attacking the people who resist their new developments, Tesla was too intelligent to waste much time answering his adversaries and critics. For the most part, he just ignored them. As a result, his scientific zeal and mental focus was not blunted by hostility and defensiveness—a point that many innovators in all fields of endeavor should take to heart.

The most significant feature of Tesla's style of working, though, was his use of the mind. He worked extremely hard to develop and train it, and applied it prodigiously. He sacrificed many of the usual enjoyments of life, including emotional relationships, in order to keep his mind clear and precisely functioning. And what a mind it was! Triggered by a finite set of observations and questions, and armed with his intense concentration and dedication, Tesla could soar into a complex set of thoughts and abstractions which laid bare the secrets of nature to him. In effect, his own mind became his true experimental laboratory, in which he could test his ideas and investigations far more perfectly than in any physical laboratory. In his mind, he could examine universal law, and apply it to practical problems. In his mind, he could consider the interaction between physical phenomena and the phenomena and forces of the other dimensions of life. Most people believe the mind is subjective and unreliable. Tesla, however, *proved* that the mind is the one true objective laboratory for scientific investigation.

As stunning as it may seem, Tesla actually created and *built* prototypes of his inventions in his mind. He generated not only good ideas, but also working models of them. He could switch on these models, let them run for a month or two, then tear them down and inspect them for wear—all in his mind! When he then built a physical duplicate of the mental machine, it would work perfectly the first time he turned it on. No scientist has ever been able to work that well in a physical laboratory.

It might be said that Tesla was a scientific transcendentalist, but it may be more nearly correct to label him a transcendental scientist. He transcended not only the scientific thought of his day, but also transcended the material world and learned to investigate all aspects of life associated with any set of phenomena. Beyond a doubt, this capacity to rise up above the level of the obvious was his greatest contribution of all—if only we open our eyes to see it for what it is.

I am shamelessly fond of the way that Nikola Tesla worked. I believe it to be an outstanding example which can help every intelligent person who seeks to extend his understanding of himself, his work, and life in general. To my way of thinking, Nikola Tesla gave the word "genius" a new depth of meaning and potential.

It is for this reason that the interview which follows is devoted primarily to a discussion of the nature of thinking and the training of the mind. Some scientists may be disappointed that there is very little technical commentary; they may wonder why Tesla was not asked to provide the technological guidance needed to reproduce his death ray, to solve the energy crisis, or to reveal the hidden mysteries of physical life. Indeed, some may conclude that because these elements are missing, the interview is not genuine.

Suffice it to say that the purpose of this interview, as well as all the other interviews in this series, is to encourage the intelligent people of the world to develop their own minds and their own creative genius. Each spirit who participated was chosen because of his or her ability to discuss some aspect of the development of the human mind or the human inspiration. Moreover, the choice of subject matter for each interview was essentially left to the spirits themselves. In other words, in this interview, it was Tesla himself who chose the major themes to cover.

For example, he talks extensively about scientific humanism—the need for scientists to concern themselves with the welfare of civilization more than with naked scientific truth. In this context, Tesla explains that many of his inventions—such as the death ray—were never fully developed because they would have been harmful to mankind. He mysteriously implies that it will do little good to try to reproduce them, as there are obscure but powerful forces which will confuse and confound such efforts. His comments seem to indicate the existence of benevolent powers who have far more influence on the progress and protection of civilization than most people suspect.

On the other hand, Tesla does indicate it would be useful for scientists to pursue the applications of his work which would be more beneficial. Working with the earth's magnetic field and broadcasting electricity through the earth's atmosphere were two areas he seems to approve as fruitful. He also makes some suggestions for solving the energy crisis. But a major point he stresses throughout these discussions is that approaching these problems in his *style* of working is the real key to understanding his ideas and principles.

In response to a question, Tesla demonstrates how a knowledge of electricity can also contribute to the healing of the physical body. Apparently, the grasp of electricity he attained is virtually unbounded. Having penetrated to the very essence of the principles of electricity, he is able to recognize the application of those principles to a vast array of electrical phenomena not directly related to lights, motors, or machines. He therefore discusses at some length the subtle electrical patterns and vibrations that can affect the physical body—favorably and unfavorably. For those willing to leap mentally to a level where they can understand them, he presents a number of intriguing hints.

Perhaps the most fascinating area of discussion is Tesla's commentary on the nature of thinking. He takes considerable pains to describe the fifth-dimensional nature of thought and the power of ideas. Since the term "fifth-dimensional" may be unfamiliar to many readers, it may be useful to describe its usage in this context. A dimension is a potential for movement. In one dimension, movement is linear. In two dimensions, movement can occur through a plane. In three dimensions, movement can be through the height, width, and depth of solids. In four dimensions, the movement of "projection" is added. This can be projection through time, as in the growth of a tree, or projection through space, as in the intuitive ability to be aware of the feelings and moods of another person. In five dimensions, the added movement might be called "simultaneity." It can be seen in the growth of a whole species, or in the fact that a single archetypal idea can have thousands of different, simultaneous applications. A further explanation of the fifth-dimensional nature of thinking can be found in an essay I wrote a few years ago with Carl Japikse, "*The Mind and Its Uses.*"

Definitions of some of the technical, occult terminology used by Tesla in the interview can be found in the glossary at the end of the book. I have made no effort to define standard scientific terms, however.

For those readers who find Tesla as fascinating as I do, I would highly recommend the biography of his life written by John J. O'Neill, *Prodigal Genius*. A science writer, O'Neill knew Tesla personally. His book provides marvelous insights into the personality and mind of a remarkable individual.

Unlike the other interviews in the *From Heaven to Earth* series, in this one I do not ask the questions. Instead, I am the medium. It was felt that David Johnson, who acted as medium in the other interviews, did not have enough of a scientific background to allow Tesla to express himself clearly. While I certainly do not have a background in electrical engineering, my training as a medical physician did make me a little better prepared for this particular interview.

The questioning is conducted by Carl Japikse, who is the publisher of this series; David; and a friend whom we shall call Richard Gordon. Mr. Gordon is an engineer.

Tesla: I would like to make a statement on the nature of this particular sitting, because we are using a different form of [mediumistic] communication than before, since Robert is the medium. He has different mediumistic skills than David or other mediums. In essence, Robert's consciousness is on two levels. He's here with you and he is also in another dimension, where I am. We are "talking" together, if I may put it that way, in this other dimension and sharing his subconscious facilities at your level. It's a kind of cooperation, really. I think it's a pretty comfortable arrangement.

Japikse: Is this a higher form of cooperation than usually occurs in mediumship?

Tesla: I don't know.

Japikse: Can you describe this level where you and Bob are communicating? Is it your normal level of focus as a spirit?

Tesla: Not quite, but it is comfortable for both of us. In your terminology, we are creating a synergy of ideas and are very free to move about quickly—not just in a spatial sense, but to *move inside ideas.*

Japikse: So, it's not just the medium who changes his state of consciousness to enter a mediumistic rapport. The spirit who wishes to communicate must also change his state of consciousness, at least to a small degree. Have I understood that correctly?

Tesla: That's true for me, at any rate. I really cannot speak for other spirits.

I'm told that there are not many mediums like Robert around—not many who can be conscious in two dimensions at once and transmit ideas clearly.

Japikse: I have noticed that when I communicate clairaudiently with a spirit, such as I've done with you before, there is a real quickening and stimulation of the thought processes when our two minds come together. Do you experience the same kind of quickening as a spirit when your mind is *en rapport* with the mind of a physical person?

Tesla: I take it you are referring to the rush of force which comes with contacting a higher level of thought. It's the force of ideas or thought that seems to stimulate you, as they are "stepped-down" to the physical plane. I'm finding this terminology very unsatisfactory, because the plane of thoughts really isn't "up there" and the physical plane "down here." The dimension of thoughts interpenetrates the physical plane. Thoughts are right here spatially, but in another dimension, too. Words are a little inadequate.

Anyway, to answer your question, no, I don't experience a big rush or thrill, because I *live* with that force of ideas. Maybe I should say I live in the great force of thought. And it's a very comfortable place for me.

To put this in ideas akin to the work I did physically, whenever a connection is made between two poles, a current will flow between them. If the current is electricity, it will flow either one way or the other. But when you are dealing with the higher dimensions, and the current is thought, it can flow both ways simultaneously. In a very peculiar fashion, which I don't think I would be able to describe, there is a transference which is more than a two-way pull between the one pole in the higher dimension and the receiving pole in the earth dimension—the receiving mind.

Japikse: That's very interesting. As I was thinking of questions to ask you, it occurred to me that the work you did with electricity was quite symbolic of our theme in these interviews—from heaven to earth. It strikes me now that what you've just been describing is very much along the same line: the flow of energy from one charged area to another, which becomes charged as a result. Since today is the full moon, would you have anything to say about the relationship between the phases of the moon and the concentration of energy? Is this an area scientists ought to be investigating more fully?

Tesla: Of course they should, although it might prove to be a lethal strain for some scientists I've met. *[Laughter.]* Yes, investigating the phenomena associated with the full moon would be very worthwhile, because it would lead scientists to study and, I would hope, comprehend the subtle planes of matter. I will have to use occult terminology, but I am talking about the etheric subplanes of the physical dimension and the phenomena and substance of the astral plane.

When it is full moon, as Robert likes to say, it's high tide not only on the oceans but also on the subtle planes. The subtle body of the planet is affected—etherically, astrally, and even mentally. It is high tide, which simply means that matter is stirred up. The electrical and magnetic forces are stirred up. In fact, anything that has to do with matter, whether it is dense physical matter or the subtle matter of the inner planes, is intensified.

Of course, I am *not* suggesting that the speed of light is faster at high tide, but there are some measurements which could be made, using current technology, which would demonstrate that there are some slight but amazing variations in electromagnetic phenomena. And I'm not referring to increases in static—although, if people would look at what sometimes is thought of as electromagnetic static or interference, they might be surprised by what's going on.

Yes, electromagnetic phenomena are enhanced at the time of the full moon. That is why the emotions are intensified then. As you know, emotions carry a kind of magnetic charge. It is common knowledge that the emotions are more stirred up at high tide, the full moon. In point of fact, there are a variety of electrical and magnetic reactions which are intensified at that time. This phenomenon could most readily be isolated and studied in plant life. It involves subtle electrical changes concerned with chemical and especially biochemical reactions. These reactions are extremely sensitive to subtle changes in electrical fields, and the most sensitive instrument for detecting these changes would therefore be living matter, especially plant life. This is because living matter, whether it is animal, human, or plant, is always multidimensional. Any living organism has not only a physical body but also an etheric and astral body. In the higher animals and humans, there is also at least a slight trace of a mental body, although I sometimes wonder about some people. *[Laughter.]* All of these bodies or dimensions interact, producing manifestations such as growth, health or disease, or movement. These manifestations are all affected by any change in the electromagnetic field. And so they are definitely affected by the high tide, the full moon.

Japikse: An interested scientist, then, could actually measure a reflection of the growth factor in plants by experimenting with the electrical sensitivity in those plants and how it is affected by the tides.

Tesla: Sure. And a good place to start such an investigation would be with the alleged "folk tales" about planting such and such a type of plant during the dark of the moon—that is, the new moon—and planting others at different phases. There are also rules for when to prune, and so on. That folklore, I'm told, is based on centuries and centuries of commonsense observations by simple but intelligent people who patiently noticed what happened when they did *this* and what didn't happen when they did *that*. They noted what went wrong and what went right. I would call them the original scientists. They actually *observed reality*. They drew logical, sensible, rational, and empirical conclusions from those observations, tested them by using them, and then passed on this knowledge to succeeding generations. Obviously, such a body of knowledge is easily contaminated by prejudice and superstition, but there are still a lot of good ideas preserved in that fashion.

Now, the value of this folklore to a scientist is not that it will help him learn when to plant his garden or when to pull weeds, but that it provides an opportunity to investigate the relationship between the physical plane and nonphysical dimensions. The physical plane is surrounded by and constantly influenced by subtle, invisible dimensions of energy and substance. And the basic connecting link from one dimension to another, from one grade of substance to another, is the electromagnetic forces in life. There is a profound need for new understanding of the essential physics of the universe, the essential nature of the universe, and the essential electromagnetic nature of life itself and all life forms.

Japikse: Many scientists hearing you talk about folklore would say, "Oh, but that's not logical, that's not scientific." What is the true scientific attitude?

Tesla: I am tempted to make a remark that the minds of some scientists really belong in ants, but somehow they got mixed up in the wrong species. But I won't say that. *[Laughter.]*

Let me take some liberties in answering your specific question. The major thing I'd like to do in this interview is set down my philosophy of science and scientific investigation. I think the example of scientific investigation I set during my physical life was probably more important than any specific inventions or developments I was responsible for. That's what I would like to inspire people to understand in this conversation.

I had the capacity to grasp the broad fundamental essence of electricity and other phenomena and convert these theories and fundamental principles into concrete, practical devices and machines. This is a way of using the mind that is distressingly deficient in the modern scientific scene. There are too many scientists who are experts in facts but do not understand what they are doing with them! They are specialized to the point of absurdity and have lost sight of the roots of the study they are pursuing. These scientists are well grounded in concrete intellectual theory, but in many cases this is just a crude intellectual construct which does not bear a very close relationship with reality. They fail to *think* about these things—why, they don't even *speculate* about them!

If there is a relationship between folklore and a certain scientific theory, how can it be illogical to explore that relationship? I say it is illogical *not* to explore it!

It is all right to accept these intellectual constructs for lack of something better, but scientists should not accept them as though they were inviolable. They should not accept them in such a way that they limit the scope of scientific investigation and cloud the scientist's capacity to *observe life!*

There are a number of new breakthroughs in science which are urgently needed at this time. We on the inner planes are ready to proceed, but these breakthroughs won't be able to come until we have a different sort of consciousness on the scientific scene, able to behold the broadest possible range of a subject.

I suppose I am describing the ideal of a more humanistic scientist—an investigator, scientist, or engineer who understands that what he is working on is also significantly related to civilization and humanity. Such people do exist today; they have always existed. But this element is sadly missing in most scientists today. The wholeness and humanism of science is not being taught or emphasized. Some people may give it lip service, but they don't really live it. This is very sad, because the real stimulus for scientific discovery is the collective needs of humanity. Whenever a scientist effectively focuses this collective need through his work, great things can be called forth, depending, of course, on the current state of knowledge and technology. The more a scientist can grasp the needs of society and humanity, the more he will find the energy, stimulus, and impulse to make great breakthroughs in chemistry, biophysics, or what have you. This is a very important point—and one that is all too often thrust aside as "philosophical ravings." Yet I am talking about harnessing the energy that is required to stimulate new discoveries. The *energy!*

This is the approach I used in my life, so I know it works. And if the scientist can behold the applications of his intent, based on the needs of humanity, then he will understand how he must proceed to get his breakthrough. A scientist discovering new theories, new ideas, new techniques, or new technology must be thought of as being a midwife—no, that is the wrong term.

Dave Johnson: A mother?

Tesla: I was looking for another term besides mother, but I'll accept that. The scientist is the mother for new thought children. The human consciousness on the physical plane is designed to serve this role. The inspiration for new ideas comes from beyond the physical level of consciousness; the mind receives this inspiration and nurtures its physical unfoldment. It is the womb of new discoveries.

Johnson: Where does the scientist start? With theories or with phenomena?

Tesla: With the observable phenomena, of course. That's the trigger device that would stimulate the scientist's curiosity. But an investigation should be based on something more than observed phenomena alone. I don't want to say it should be based on a theory, because it's not a theory. Scientific investigations should always be based on an understanding of the fundamental essence of things. A theory is a human construct; what I am saying is that the humanistic scientist will always base his investigations on an awareness of the patterns of life. Unless the scientist connects himself with the essence of life before he proceeds, he will never achieve any really significant breakthroughs. He won't be connected with the energy of discovery. The "scientist" who starts with theories and then tries to prove them is dealing only with a hash of ideas. He's trying to take the results of this and the effects of that and the by-products of something else and put them all together and create something meaningful. It's a good way to cook up leftovers, but it doesn't lead to scientific breakthroughs. *[Laughter.]*

Johnson: The scientist, then, should have at least an understanding of observable phenomena to start from.

Tesla: Yes, inspiration alone is not enough. The scientist must prepare himself by diligently studying all the known science of his day and age, filling his mind with facts and practical experience and mathematics. You cannot be a good mother unless you have acquired all the parts needed to generate and deliver the new thought child.

Johnson: I'm not a scientist myself, as you know, but I have observed any number of "scientists" who start with a pet theory and pretty much avoid the actual phenomena. That's always seemed like backwards science to me.

Tesla: It is. That's why I didn't want to use the word "theory" a few moments ago. A theory is an intellectual construct, based on the scientist's interpretation—sometimes his feeling—of what the observable facts and phenomena mean. The good scientist starts not with theories but with an awareness of the essence of life, and works from there. In dealing with physics, electricity, and similar phenomena, for instance, the scientist would want to start with an understanding of the essence of the physical plane. In the nineteenth century, this essence was called "the ethers." It wasn't too well understood then and is even less well understood today, much to our disgust. But it's in the ethers, or the etheric plane, to use the occult terminology, where I found the roots of all the phenomena and potentials that I found so fascinating and worked with in my lifetime. It is the operation of electrical and physical phenomena at these etheric levels which must be studied, not somebody's theories, if the scientist hopes to accomplish anything significant here on the earth plane.

That's not to say you ignore theories. The good scientist reads about the theories of the day, of course, and can be inspired by them. It's a little like reading a novel and being inspired by some of the ideas in it. The novel does not necessarily contain the ideas as such, but reading it does trigger insights of your own. Other peoples' theories are useful in that same way.

The tragedy is that scientists get stuck with a pet theory and their minds stop and they can't behold anything new, anything beyond the theory. I think that sort of one-channel "thinking" has been called "vertical thinking" by someone. If someone is working on transmitting electricity, for example, and gets stuck on the idea that it must be transmitted with 60-cycle current and X number of volts, that can be a great limitation. He will only be able to think within the framework of short distances. With lateral thinking, however, he would realize that he could use high frequency currents or extremely high voltages and make the long distance transmission of electricity quite possible.

One-channel thinking could almost be called "constipated thinking." It's a great hindrance to science. The people who work with a pet theory in this way are just digging the hole deeper—without realizing that there is nothing worthwhile to find in the direction they're digging. They need to look in new areas.

Never confuse the role of technician with the role of scientist. Technicians like to measure things and record data and accumulate a hoard of statistics that they then try to interpret. Careful measurement of data certainly has its usefulness, but a scientist is an individual who is motivated by a spirit of inquiry to discover the way a certain body of phenomena behaves, how it can be manipulated and modified, and how it can be harnessed for some practical application. He learns to manipulate the phenomena so he can duplicate or recreate them. You see, there needs to be at least a semi-practical end point to this scientific inquiry. I know that idea will upset a lot of people, but it is vital to keep it in mind. Too many people seem to think that being a scientist means collecting and observing facts. I say there must always be a practical end in mind, and that practical end is best defined by the scientist's constant re-appraisal. He should ask himself: what does this phenomenon mean? How can I use it? What good would it do for me or anyone else to understand why this happens or how this works? By asking questions like these, the scientist helps ground the mental energies of this knowledge in the physical plane.

I want to digress a moment to talk about the nature of ideas. A lot of what people call "ideas" are just fudged-up emotional messes, but that kind of idea is not going to help anyone solve problems. True ideas are concepts. They are fifth-dimensional and in constant motion. That means that an idea cannot be observed or dealt with the same way you would observe or deal with a stone. To deal effectively with ideas, the scientist needs to observe a variety of interrelated phenomena and extract from those observations the repeated patterns, principles, parallels, applications, and so on. In the process of observing these phenomena in this very practical, sensible, and rational way, it is possible for the mind of the scientist to "reach up" and grab the very *essence* of the idea which is manifested through them. To grasp this essence, of course, he has to enter the plane of thoughts, the mental plane. To put this allegorically, he reaches out with two or three or ten "thought fingers" and grasps this fifth-dimensional idea, which seems to be whirling in space with about 120 corners.

Japikse [laughing]: That's *not* a literal number, is it?

Tesla: Well, I made it up, but it's not too far from accurate. Anyway, the scientist brings three or four of these "corners" of the idea into his consciousness. Now, to be able to do that requires a capacity for thinking abstractly and an ability to project, apply, and associate the essence of the thought to concrete knowledge and phenomena on the physical plane. That's how a fifth-dimensional force—the real idea—gets delivered into the physical consciousness of any intelligent person, not just scientists. It's a very important part of true logical thought, but to understand what I am saying, you have to realize that logic involves an inner-dimensional activity. Logic has more levels to it than most people understand. One can talk about linear logic, that one plus one equals two and two plus two equals four. This kind of logic is based on a progression, where you add brick to brick to brick and end up with a wall. But there's another type of logic—another aspect of logic, really—where you learn to think in somewhat different terms. You observe your three bricks and understand they can be used to make a wall, but then your mind leaps to another level and you realize that this wall can be part of a fence, or part of an apartment house, or part of some other structure. Or they can be laid into a sidewalk. You are tuning into the essence of bricks—the purpose of bricks. This can be done, but only by someone who is able to leap up to a mental level where he can appreciate that bricks have greater bricks, have greater bricks, have greater bricks, just as fleas have lesser fleas, that have lesser fleas, that have lesser fleas, that have lesser fleas. *[Laughter.]*

As you move into the more subtle dimensions of thought, you move from concrete, observable facts and phenomena to a more generic level of thinking, where you deal with the archetypal essence of the phenomena—the patterns of thought, the principles, and, I daresay, the generalities. You begin to deal with the principles of natural law. From there, you can actually tap into the very essence of concepts—what is truly the archetypal essence of ideas. At that point, you begin to synthesize the different levels of thought and can then project that synthesis into all kinds of experimental situations. It's like being able to say to yourself, "If I understand the principles, patterns, and essence behind *this* particular phenomenon of electricity—or whatever—then I can apply those same principles, patterns, and essence in all these other ways, too. To do this and that and something else over here." And, indeed, you *can!*

But this type of thinking does require the ability to climb into a higher dimension, where you are dealing with the very center, the very focus, the nexus of the idea.

Japikse: Let me make sure I understand this. An idea

figuratively has 120 corners. As I examine some phenomenon or phenomena associated with it, I am probably grasping only three or four of those corners. But just as the three or four bricks imply the whole wall, I must remember that the three or four corners of this idea imply the whole idea, the synthesis of the idea. To work logically with this idea, I must never take the three or four corners I'm dealing with out of the context of the full 120 corners—the other applications or uses for the essence of the idea. Is that part of what you are saying?

Tesla: Well, it's at least three or four corners of what I am saying. *[Laughter.]* Yes, I think that's a fair statement. And I have just been inspired with an analogy.

If you will, think of an ordinary bicycle wheel. Now, a bicycle wheel has God only knows how many dozen little wire spokes. At any given moment, however, only one point on the circumference of the wheel is touching the ground and it looks as though the weight of the bicycle is being borne entirely by just two or three spokes. It isn't, of course, but an "amateur observer" might think, "We can cut away all the spokes except the most vertical." *[Laughter.]* Many thinkers "think" in that way. They assume they have a complete idea because they have registered two or three points of the idea. Actually, they are only looking at the circumference of the idea and the few spokes closest to them. If they would move on a bit, they would realize there are many other facets, phases, and applications of the real idea. There are many other faces of the idea to be viewed—if they start regarding the idea fifth dimensionally. Doing this is like moving the bicycle wheel just six inches forward. New spokes descend to earth to support a whole new body of information.

Japikse: It seems to me that the ability to think in this way is what distinguished your career from the work of other scientists. Most people find it very difficult to understand how you were able to continually bring through new ideas, and in the course of a few months accomplish what others would consider the work of a whole lifetime. I presume it was because you had developed your ability to work fifth dimensionally, with all the spokes on the wheel.

Tesla: Yes. I was able to work that quickly because I was a genius. That's important, you know. Men and women should learn to appreciate and honor more of their good side and really strive to excel. They should strive to develop a well-functioning mind and then use it in disciplined work. I am very proud of the fact that I was a genius and manifested that genius in many ways. I would hope that others might study what record there is of my life, not so much to learn the secrets of my inventions—that's really not all that important—but to learn my manner of thinking and approach to the scientific method. I'm not saying people should try to imitate me, because the imitation of genius is not genius itself. But if I could think fifth dimensionally, others can, too. The mental processes of my genius can be duplicated.

In physical life, I was what might be called a profoundly intellectual person who often neglected common habits and who didn't think too much of the standard aspirations of the average person. I often completely lost myself in my intellectual pursuits and, frankly, did become drunk on the force of certain ideas and trains of thought and speculation. It is a risk any deep thinker would take.

The fact that I was able to focus into the physical plane certain inventions and ideas indicated that I had myself under control and could pull myself away from my fascinating intellectual pursuits enough to bring them back into the everyday world and put them to use. That's part of the scientific method, too. I was a very disciplined person, disciplined mentally and physically. I had some rather rigid personal habits—indeed, it was a rather peculiar personality, in retrospect—but I used the body and personality as an instrument for my mental workings. It was, therefore, very important for me to have a tightly disciplined personality and mind. Otherwise, the very force of the ideas I dealt with would have completely unbalanced me. It would be wise for people to understand this aspect of genius.

To borrow some electrical analogies, if you're going to store or transmit a very great charge of electricity, your electrical hardware must be extremely well made, well-insulated and grounded, or everything can explode in one blinding flash of electrical discharge. The force can burn out the circuits and destroy everything. In a similar fashion, the human mind is an instrument which is designed to carry an exceedingly high voltage [fifth-dimensional ideas], *but only when it is properly prepared with a sensible, mature personality!* Only when it has been trained in logic and careful observation, only when it has been trained in intellectual integrity, only when it has been trained to be pragmatic and empirical, is the personality then prepared to be a fitting vehicle to safely carry the tremendous force of real ideas.

This may not make a whole lot of sense to people who are used to thinking of thinking as the *manipulation* of ideas, but I don't consider that genuine thinking. They end up primarily manipulating *memories*—perceptions they have registered, speculations they have entertained. They are registering only the sensory input of what their eyes have seen and what their ears have heard. There is no force or power to it, so they wouldn't be aware of the need for preparing the personality. But when I talk about thinking, I'm talking about a process that connects you with the abstract essence of ideas, and that can be extremely powerful.

The abstract essence of an idea can only be contacted by a skillful, well-prepared mind. It is not contacted through observation or speculation alone—it is *summoned* from heaven by the intelligent and skillful focusing of an enlightened mind.

Johnson: I suppose the capacity to think in this way would be of immense service to the humanistic scientist?

Tesla: Oh, very much so.

Johnson: I have been interested for many years in the humanistic man. Would you care to talk a little bit more about the humanistic scientist?

Tesla: Well, the humanistic scientist is an individual who deals with nature in all its diverse aspects, not just the laws of physics or gravity or mechanics. He deals with the *whole* of the universe, knowing that all of the manifest parts and laws and phenomena of this universe are somehow related. The whole is related to itself and all of its parts. And humanity as a whole is one of these parts, one of these parts of the whole.

In dealing with electricity, for example, the humanistic scientist keeps in mind that the scope of electricity is much broader than the one small application he is focusing on. There are natural electrical phenomena that keep your physical body alive and all life forms alive. The whole planet is charged electrically. But every natural use of electricity serves a purpose in supporting the manifestation of life. To put it poetically, God *uses* electricity, He doesn't just study it! So the humanistic scientist tries to do the same. To him, electricity serves no purpose unless it can be put to work for the benefit of civilized humanity. I'm talking about applied electrical phenomena, of course.

In my lifetime, there were hundreds of ideas, inventions, and creations which never escaped my mind. I never set them down on paper, patented them, or made any working models of them. This might have seemed a bitter frustration to me, in light of what I just said, but this is the opposite side of the coin. I realized that many of these ideas had a destructive potential vastly greater than the hydrogen bomb. And they would have been easy enough to assemble that some monstrous person working in his basement could have created a device that would destroy high-rise buildings and cause a great deal of havoc. It would have been irresponsible to loose *those* ideas on humanity. The humanistic scientist is aware of those concerns as well.

I will make an aside here. The so-called "Nine Unknown" have something to do with policing this department.

Johnson: That's good to hear. Did they have something to do with the disappearance of the "death ray" you supposedly developed?

Tesla: That didn't exactly disappear. As you know, I had the sort of mind that could visualize an invention or idea in great detail. I could say I had a photographic memory, but it wasn't quite that. I could build an invention in my mind. So, there was no need to write down blueprints or diagrams. My own mental memory files were much more complete and efficient than written notes would be.

Johnson: Oh, I see—the only existing model of that particular invention was in your mind. It was never built.

Tesla: Basically so. By the time I came up with that idea, there was no question in my mind that it would work. After all, when I discovered the principle for generating alternating current, I worked out all the details in my mind. I literally visualized the generators right out here *[pointing to the air space at arm's length in front of his face]* and mentally built them, operated them, and refined them until the design worked. I even tested their efficiency—all in my mind! I could set them in operation and let them run in my mind for a week or a month and then dismantle them, to determine the site and degree of wear and tear. That way, when I finally did build the physical generator, it always worked right the very first time I turned it on. So there was little need for me to write down notes about the "death ray," as you are calling it. I did talk about it, but I spoke in the most guarded terms, and I never explained the basis of how it worked. It's a very simple principle of molecular disruption. I'm amazed that people haven't come across it, but I won't talk about it.

Johnson: I have a question which may sound silly, but I've always wondered if it would be possible for someone to develop a device that would actually photograph the thinking of another person.

Tesla: Of course.

Johnson: Is this something we will see in our lifetime?

Tesla: Of course. I'm surprised someone hasn't done it already. It's utterly simple.

Johnson: I thought it might be something like the use of computers to rotate the diagram of an object three dimensionally.

Tesla: It will probably take advantage of a device called the image intensifier. It's used in television.

Let me digress a moment. Some of the other spirits participating in this project have made the comment that when you search into phenomena—whether it's the phenomena of physics, music, art, human physiology, or whatever—and get beyond the physical forms, energies, and activities, you eventually arrive at the inner essence. Then, behind this essence is something even deeper or more fundamental, more profound—the Father. You find that everything is the handiwork of the Creative Force of the whole universe, and that one of the primary laws of creation is order. It is an orderly universe. Everything hangs together and works together, somehow. Life appears chaotic only to the inexperienced eye or ear—the eye that does not see and the ear that does not hear. The universe is orderly—the whole of Creation, really, is a beautiful mathematical creation. I would call it a mathematical or geometric creation. Someone more poetic might call it a musical creation.

My point is that there are harmonies and rhythms which permeate all of Creation. This idea is as fundamental to ordinary mathematics as it is to electricity. There are octaves of energy, definite waves and rhythms that can be measured, frequencies and amplitudes, and so on. From these simple elements are produced an almost unlimited number of variations. You can go from the very subtle to the very dense in this way—from pure energy to a dense physical form you can touch and move around. I hope that's not too confusing.

Johnson: No, it's very clear.

Tesla: Then let's apply those ideas to the notion of measuring and photographing or mapping the human subconscious or thoughts. Because there are various octaves of energy in creation, there are subtle counterparts to everything existing in the physical octave. Well, it's wrong to say the physical forms have subtle counterparts; it's really the other way around. The essence is here *[pointing above his head]* and it has a lower octave down here *[lowering his hand to the waist]* as some variety of subtle energy, and then there is another octave lower yet which is still energy but more gross—perhaps even detectable by crude physical voltmeters. And yet there is an even lower octave—the octave of physical forms, mass, and matter. All these levels are octaves of the same essence. If you understand the essence, then you can understand everything below that and how the octaves interact. For each octave, there is a certain pattern of amplitude and frequency. There will be millions of variations. Understanding this idea will give you a clue as to how to map human thoughts.

One extra point needs to be mentioned. By applying a charge of external energy to a relatively closed system, you can selectively energize a given octave of energy. This is a simple process I used all my life, yet still only a few people have understood this subject. It's the basic principle of resonance. By selectively applying a specific vibration, you strike a resonance in one of these subtle bands of energy, and this then stimulates the lower octave, which stimulates a lower octave yet, until a simulation of the subtle energy of the higher octave—normally invisible to the human eye—becomes visible.

This is what happens in Kirlian photography, although the energy is only being stepped down one level. A certain type of energy is applied to an aspect of the etheric energies, although it's not exactly what the scientists think it is. It's something else, but it stimulates the etheric energies so they can be photographed.

I'm not sure I'm making myself clear, and I'm a little annoyed, because this is as simple as one plus two equals three. It's that simple. If it's not clear, please ask questions.

Johnson: No, I understand.

Tesla: Let me try to use occult terms. If you stimulate the low astral, that will produce a corresponding effect in the upper etheric planes of matter. You could also then stimulate the upper etheric planes to produce an effect on the denser physical plane. That effect could be a photograph. I hope that will be clear to some.

Japikse: These are all fascinating ideas. I'm wondering if some of our readers might find it helpful if you would talk a little bit more about what you mean by the essence of ideas. Could you give an example of this essence? Or at least an analogy?

Tesla: Okay, but I'd like to switch from the field of electricity to the field of psychology. The readers might be able to relate to that a little more easily.

Japikse: Well, I've always thought the essence of electricity was pretty much the same as the essence of psychology.

Tesla: Exactly. Now, it is possible to look at human nature as it manifests itself in behavior and personal relationships. You can observe all kinds of variations in that behavior—for example, how someone treats a policeman differently than a priest, or a car salesman, or a waitress who sells him coffee and donuts. Or, you can observe the habits, ambitions, and self-images of people. These are the phenomena of human nature. They can be observed, but you can also go beyond observation and begin speculating on why these people behave in those very different ways. As you do, you have to appreciate that there are some profound inner dimensions to human consciousness. They are real, even if they are not well understood. Eventually, as a result of these speculations, you begin to suspect that there truly is an inner seed of character in every human being—a tremendous inheritance of good and bad, of potential and disease that influences people from deeper dimensions. This is a subtle but powerful and persistent influence on human behavior.

As you speculate in this way, your mind begins to leap up to some of these more subtle levels of thought. You come close to the essence of the idea you're pursuing, until you actually begin to summon the very presence of your own soul. The soul, of course, is not just an idea, but it lives on the level of pure ideas. It begins to impress upon your intelligent, speculating mind the reality of its presence. It nurtures and stimulates a train of thinking which, if sufficiently pursued, will lead to some degree of realization about the inner nature of men and women—that they are spiritual beings and that all the phenomena of human behavior can be traced back to a spiritual core. This core is the real center of our life force, the real cause of our good ideas, our noble and loving thoughts, our involvement in life, and all the varied expressions of our humanity in daily living. When you understand that, you're dealing with the essence of human psychology—the essence of all ideas about human behavior. Does that answer your question?

Japikse: Yes, I think so, but I want to clarify it a little. Would it be fair to say that the essence of our ideas about human behavior would not be our understanding of our spirituality but really our spirituality itself?

Tesla: Our experience of the idea of our central spiritual nature would be the realization that we are spirits living in physical bodies. The *force* of that idea is our spirit, but our *experience* of that idea is a realization. We are inspired with some degree of realization, but the essence of the idea is not something we "think up." Nor is it something we necessarily see or hear or touch or taste. The realization of this essence is a mystical communion. We encounter the force of the idea and know for certain that that's the way it is, *because that's the way it is!*

Japikse: So the process of thinking, or interacting with this inner essence of ideas, is the process of realization?

Tesla: The process of thinking *could* lead to realization. Not all thinking, however, leads to realization, or even to wisdom or inspiration. In fact, not all thinking leads to knowledge. Some of it is just fooling around.

Japikse: But if the process of thinking is dealing with the essence of ideas?

Tesla: There's a good chance, then, that it will lead to realization, yes.

In one of the essays you and Robert wrote recently ["The Mind and Its Uses" in *The Art of Living* Volume IV] you did an excellent job of setting forth the notion that wisdom must be manufactured out of observations, ideas, and the logical relationship of facts and knowledge. This is the process I am trying to describe, but I am adding an extra fillip of my own. I am trying to emphasize that proper thinking literally summons from heaven the very essence it is leading toward.

Let me try to give you a visual model. If you tinker together a set of observations, ideas, and facts and if you carefully apply them, you build a pathway, a superstructure that is grounded on earth but leads to heaven. Your concrete ideas and observations cannot reach heaven themselves, but they act something like an antenna. They magnetically invoke an impression or force from higher dimensions, from heaven.

Now, the antenna of concrete thoughts does not always pick up the essence of ideas, any more than a radio antenna always picks up clear signals. Sometimes, it picks up thought static. That's why I said the process of thinking does not always lead to realization. But it can.

Japikse: Along that line, I remember reading about your experiments in Colorado, broadcasting energy throughout the whole world. The first thing you had to do was to see if the earth itself was charged electrically. If it had not been, you would have had to charge it before you could have sent energy through the atmosphere. You found that it was already charged. That strikes me as parallel to what you're saying now—that the mind must be charged and developed before it can receive inspiration from heaven.

Tesla: Yes, but you must realize that I'm talking about a higher octave than physical magnetism. I don't want anyone reading this book to hook a wire to his left ear and plug himself into the wall, thinking that if he runs an electrical current through his brain he will somehow become brilliant. *[Guffawing.]* It is true that we need to charge our conscious minds with rich ideas and observations. This is an absolutely essential step in learning to think. We must learn not just to handle observations and facts; we must also learn how to apply those facts in a hierarchical or higher sense. We must understand that our observations form certain patterns and that these patterns imply certain laws, which in turn are related to other laws and possible applications over here, and here, and here *[gesturing]*.

You have given me an opening to say something I've been

really waiting to mention, so if you don't mind, I will change the topic slightly.

Japikse: Sure.

Tesla: I'm referring to those experiments in which I broadcast electricity through air. It's important to understand that all matter is magnetized. All matter—whether it is physical, etheric, or even more subtle matter ordinarily invisible to physical eyes, physical meters, and even electron microscopes—is magnetized and can carry some form of current. Even nonferrous metals can carry a certain current, at room temperature, if you know how to detect that current. Like any other physical object, the earth carries a magnetic charge—indeed, a very heavy magnetic charge. This is because it is moving. And any magnetically charged object that moves produces some form of electrical current.

The earth is moving within its own electrical field, but it is also moving within a larger electrical field, which is the electrical field of the sun. There is enormous energy there to be tapped. We actually swim in oceans of energy.

Japikse: And that energy could be tapped feasibly, not just psychically or psychologically, but also mechanically?

Tesla: Sure.

Japikse: Was the work of your lifetime a step in a series of planned stages to help humanity learn how to tap some of these larger electrical fields?

Tesla: My work was very complicated. The main thing I was supposed to leave civilization, I did. My principal accomplishment was to perfect some of the practical uses of electricity, namely the use of alternating current. I also stimulated a few people to appreciate that even today mankind knows very little about electricity—and for that matter, very little about the nature of magnetism and the nature of physical matter. If the electrical engineers and physicists would stop looking at their textbooks and journal articles long enough to start looking at *reality*, they might discover some very obvious truths which might revolutionize their understanding about electricity, magnetism, and matter. Reality, you know, is the original source material.

They might discover, for example, that there are many different types of electricity, depending on how you generate it. And there are many different ways you can transmit it.

Japikse: Was the work you did just a one-man show, or was it part of a large program directed from the inner planes?

Tesla: Well, it was part of a larger program, of course. As the poet John Donne said, "No man is an island, entire of itself." In my case, I represented a very enlightened group of individuals, most of whom are on an invisible plane of existence. This group is interested in stimulating mental functions and the appreciation of the mind and how it can be used as an agent for discovery. It is interested in stimulating the growth of science and demonstrating how the prepared mind of the scientist can be a tremendous force for civilizing and humanizing mankind. It is interested in helping conquer the problems of humanity and the raw forces of nature, so they can be put to civilized use.

Now, it may sound as though I am spouting chapter and verse of some old claptrap dogma, but there is a large body of us here who are very dedicated to those goals. Through many incarnations, we try to demonstrate that the properly prepared mind of the humanistic scientist is a very precious thing which, when correctly used, can do an enormous amount to help humanity with its multiple problems.

The good name of science has been besmirched lately by the ignorant hordes who have invaded it, but proper science still has enormous application. I am being urgently impressed to make a statement that the true meaning of science and the ultimate role of the scientist is to be an agent for spiritualizing civilization and for enriching the minds of humanity with a proper understanding of what the mind can do.

The farthest reach of the human mind is that it can extend to heaven and comprehend—not just believe in or adore, but *comprehend*—the inner workings of God and His Creation. That may sound grandiose, but it is true, and there is a large number of dedicated people that I work with who have toiled through centuries and centuries to prove that truth is not something which is so vague and intangible that it cannot be discussed. Whatever is true is true *because it can be manifested!* It can be demonstrated in the physical plane. This is what I tried to do through the personality of Nikola Tesla. I did not consciously recognize this aspect of my work at that time, although in my later years I did become suspicious of it. Yet if anyone would look at the record of my physical life, they would see that the meaning of it far transcended the inventions and discoveries I was responsible for. I demonstrated a profound principle. I demonstrated what any ordinary genius with a brilliant mind can do.

Now, I know that sounds very unhumble, and being unhumble is not very popular these days. But I think it's worthwhile to make that statement. The readers who understand genius and the brilliance of the human mind will not be offended—or even think it unhumble.

Japikse: Yes, it might do us all good to start thinking of "genius" and "the brilliant mind" as ordinary. Or at least, as normal and desirable.

Tesla: Yes.

Richard Gordon: Now that you are a spirit, are you still working in the fields of electricity and alternating currents?

Tesla: Yes, I am still busy. There are a few physical people who have the capacity to understand the essential ideas behind the physical phenomena of electricity, and I am working with some of them. The whole group I am part of on the inner planes works with these people, nudging them in the right directions.

I would work with more physical scientists if there was a wider understanding of what I've been talking about—of the role that invisible and subtle patterns play in influencing physical phenomena. To control and master the physical plane, these invisible influences must be understood, mapped, and employed. Their laws of operation must be discerned. Devices that can measure them must be invented.

My work is exceedingly slow, because the people who think like this and who have the capacity to exercise their imagination do not receive very high priority. They are not recognized. Their thoughts are not given much credence. But that will change in time.

The other thing I do is teach. Not in a physical university—that would be quite a sight! *[Laughter.]* I teach in a sort of Electrical College on the inner planes. I instruct many people while they are out of their bodies at night. They do not remember this instruction consciously, but the information is placed in their minds and at the right time, they are able to use it.

Gordon: Are you doing anything with magnetism?

Tesla: Well, yes. Magnetism is inseparable from electricity, of course; one cannot be interested in one without knowledge of the other. They are parts of the same phenomenon, actually.

Gordon: It has been said that magnetic fields can enhance healing. Is this true?

Tesla: Oh, yes. It has to be the right type of magnetic field, though. Vibrations of various types will be of assistance to the physical body. Let me put it this way: the real physical body is basically a magnetic gridwork. I don't like that term, but it's in the subconscious here. The etheric body is like a magnetic gridwork. The etheric matter has a magnetic charge which is the real "body" holding together the dense physical, flesh body.

This is the body which acupuncturists work with, getting wonderful results, as they always have, when acupuncture is used properly. By adjusting the electrical flow or magnetic potential at various points within the body's distribution system, you can augment either health or disease. Within reason, of course.

Gordon: Can cross-magnetic fields destroy tumors?

Tesla: Yes, they could. The use of them would have to be contained and well-localized, though. Magnetic fields can certainly destroy flesh, and if properly contained and focused, they could destroy tumors, but they might destroy all the tissues, healthy as well as diseased.

In a liver, for instance, that is filled with tiny, separate nodules of tumor, it would be very difficult to focus selectively on the cancer. Perhaps with refinements which would let you separate the cancer cells from the healthy cells, by means of subtle changes in the magnetic potential of the two types of cells, you would be able to selectively attack the tumor cells wherever they were. It sounds good in theory, but in practice there would be some extraordinary limitations, I'm afraid. The magnetic potentials of the cancer cells are still a little too close to that of the normal cells. Of course, if you had a tumor localized in a specific area, it could be destroyed in that way. And it will be done and found enormously useful—particularly in areas where you cannot operate, whether it is cancer or not. This technique could be applied with a great deal of benefit in inaccessible areas, such as the brain or the eye or the spinal cord.

Gordon: Could treatment at regular intervals with a magnetic field prevent the inception of cancer in the human body?

Tesla: Not entirely, no. In some instances, yes, but to answer your question, you have to understand the causes of cancer. Bear in mind that such magnetic or electrical phenomena applied to the body would only have an effect upon the etheric body. Physical cancer does result from changes in the etheric and the dense physical body, but its causes are deeper. Any technique for the prevention of cancer, therefore, has to be directed at the *causes* of cancer, which lie in the astral or mental bodies—the patterns of the emotions and the mind. The true causes of disease are patterns which block the expression of the qualities of the soul. This blockage may occur at a far deeper level than the physical or etheric body.

Gordon: You mentioned that there are various octaves of energy in manifestation. When electricity flows on the physical plane, is there a corresponding flow of energy on other planes?

Tesla: There is always some, although ordinarily it isn't too much. There will be passage of electricity in the etheric levels of the physical plane, and a corresponding passage of energy in the low astral plane.

Perhaps I need to make a statement here for the record. When I refer to the etheric plane, I am referring to the uppermost and most subtle aspects of the physical plane. Physical particles and physical motion still exist at this level.

These movements of energy on the more subtle levels are what I was talking about in my comments for photographing human thought. By selectively energizing some of these higher octaves of thought, it would be possible to stimulate a resonance in corresponding lower octaves of physical matter, and produce a visible "record" of the thought.

Most human "thought" exists on what is called the astral plane. It doesn't exist in the physical plane. People are going to go "oooh" when they read this, but thought doesn't exist in the physical—not even in the brain. By selectively energizing the lower astral "thoughts," we could stimulate an etheric counterpart of those "thoughts," and then also a dense physical counterpart, which could be photographed.

Johnson: The doctor and I have been told a number of times that electricity is basically an astral phenomenon which has an appearance in the physical plane. Could you explain that?

Tesla: The energy units that go into making electrons and protons and neutrons and the subatomic particles are the energy units of the astral plane. We used to think of the physical atom as a little ball of something which had no parts; it was a unit unto itself. Then we discovered that the atom actually had all kinds of parts flying around inside it.

Scientists today tend to think of electrons as very *energized* little units, but they are really very *dense* physical units. Compared to astral energy, they are a rather slowed-down, kinked sort of energy, held to a relatively concrete form and location. I am speaking loosely now, perhaps, but not if you look at this whole proposition from the perspective of the astral plane. If you were to break up an electron, you would have dozens of tiny units of energy which would be the "atoms" of the astral plane.

That is part of it, but it's not the whole answer. Electricity as you know it is transmitted through the matter of the physical plane—actually, through the etheric levels of the physical plane. But an aspect of that electrical energy also travels through the astral plane.

Japikse: How about that! Electricity astral travels, too! *[Guffawing.]*

Tesla: The transmission of electricity on the physical plane involves the motion of particulate matter, electrons, and so on. On the astral plane, it involves a different spectrum—energies similar to radio, infrared, and light.

Johnson: Well, we have been told that electric lights, television, and the telephone are actually astral phenomena. I assume this is a simplification.

Tesla: They are part astral and part etheric.

Johnson: In particular, we were given the impression that the telephone was open to some of the more unpleasant elements of the—

Tesla: The low astral, yes. That is correct. The distribution system of the telephone network is basically an enormous antenna system connected to the astral plane. The network is physical but acts as an antenna for the astral. By giving your attention to the astral while you are on the phone, you tune into that department [the low astral]. Even if you aren't aware of this connection, you will be partially attuned to it.

Gordon: May I ask some technical questions?

Tesla: Of course.

Gordon: Can there be an isolated magnetic pole?

Tesla: Not actually. You could produce an apparent isolated negative or positive magnetic pole, but there would always have to be a counterpart *somewhere* else. It just wouldn't be totally apparent.

It would be easy to produce a reciprocating magnetic field that would be neutral on one end and either positive or negative on the other. By application of that, I suppose, you could obtain an isolated magnetic pole. It would be difficult, however; you would have problems with the inductance. But it could be done.

Gordon: The magnetic field of the earth is changing.

Tesla: It is always in a constant state of flux.

Gordon: It is getting less than it used to be. Is this going to have serious implications for humanity?

Tesla: Serious in the sense that it may be a sign of better things to come.

The earth's magnetic field cannot be properly thought of except in the context of the magnetic field of the whole solar system, the etheric currents emanating from the sun, and the physical motion of the planet itself. So, the important changes in physical phenomena are very subtle, but eventually they may be measured.

I wouldn't consider any of these changes "serious" except that many people are going to be seriously disturbed by the prospect of them. This disturbance will force them to give up some of their erroneous thinking about the universe and look a little deeper. They will have to give up "cherished facts" which aren't really facts at all.

Gordon: As the earth's magnetic field decreases, will there be increasing psychic activities?

Tesla: Not necessarily. I am told that there will be increasing psychic activity, but not due to decreases in the magnetic field. Psychic development is a separate phenomenon that's tied in with the evolution of the human race.

Johnson: There's been a lot of talk recently about a shift in the earth's axis. Is that happening?

Tesla: You're talking about the physical axis—not the magnetic poles?

Johnson: Yes.

Tesla: Oh, that dreary prediction! *[Laughter.]* I will make a simple statement of fact. The earth as a civilized community will continue to exist in great glory, and there is no need to fear awesome destruction. That is not going to happen.

I will drop an odd rock in your pond, to stimulate your

curiosity. It *could* happen, but most likely it won't. And it's not up to any specific group, no matter how odd they are, to decide whether it will or whether it won't. *[Laughter.]* It has to do with the state of the evolution of humanity and civilization.

Japikse: If you don't mind, I'd like to shift gears and ask about the type of preparation that enabled you to become a genius. A brilliant mind such as yours is not just the product of childhood, is it? Undoubtedly, many lifetimes were spent in preparation and training. Can you shed some light on what goes into becoming a genius?

Tesla: You are quite right that all talents and abilities require enormous preparation. Watching a Horowitz or a Heifitz perform, you may think it looks effortless, but you know very well it is not. Even a musical artist who is born with great talent must spend hours and hours every day in arduous practice to harness his natural talent. Manifesting genius requires enormous work, practice, and discipline.

During my physical life, I went through various austerities which forced me to put more of my attention on the development of the mind and personal discipline. I also took it as a challenge when various stupid people repeatedly told me that something "just can't be done" or "no one understands this." That was like putting a spear in my side. So I ground away—thinking, speculating, planning, and experimenting—and over a period of years, I developed an incredible ability of concentration.

I was also somewhat clairvoyant, although not in the ordinary sense. Whatever I thought about with great intensity, I could see. I could see my own thoughts. That's why I could tinker together an invention right in my mind, without needing a laboratory to experiment in. *I could see the tangible reality of my thoughts!*

I know you are waiting to hear about my past lives, but I want to stress a simple point people often forget. A person could have a thousand lives as a famous scientist and still arrive here on earth as an absolute idiot. All that potential is contained in another dimension, you see. If the child were so unfortunate to be born into a family of screwball hippies or to go to a public school where the emphasis is put on "being yourself" instead of developing the mind, all that potential might go undeveloped. The potential can only be summoned by nurturing proper habits of thinking and discipline and by responding effectively to challenges, problems, and role models. For genius to grow, an enlightened, enriched, and disciplined mind and personality must be constructed during childhood. You could import the most expensive tulip bulbs from Holland, but they will never grow into beautiful tulips unless you plant them in enriched soil at the proper season, cultivate them, and nurture them. The same principle applies to human potential.

To my mind, the most important thing I brought to earth was a ferocious aspiration and a ferocious simplicity of thought. Once I reached a certain point in growing up, these characteristics let me take over the process of developing a mature mind and a disciplined consciousness. I finished the teaching and preparation process myself.

Timing was important, too. When I started my career, there was an enormous and urgent need in civilization for a better understanding of the new phenomena of applied electricity, and a cheaper means of generating and transmitting it. At that time, electricity was more a fascinating laboratory phenomenon than a practical tool for use by humanity. The need to apply electrical phenomena was a great stimulus to me. I didn't consciously understand why at the time, but I was greatly fascinated by the prospects.

I mentioned earlier that I now teach at a kind of Electrical College on the inner planes. Well, when I was in the physical, I was a student there myself, and that, too, was part of my preparation. Every night for a long period of time I would leave my body while I was asleep and go to classes on the inner planes. I would participate in experiments in actual laboratories, attend lectures which would add to my understanding of the phenomena and principles of electricity, and then come back to earth and wake up. I often did not have an immediate recollection of what I had done, but I knew even then that I had an enriched ability to make sense of problems. I had answers to questions that had been on my mind the previous day. I had new perspectives, new solutions, and new projects to work on.

This inner planes activity is not just ordinary dreaming. There's a different quality to it. I suppose it's enriched intuition which is invoked and registered by your need to know something for use here on the physical plane. I would describe it as a rather intensive set of activities that I experienced when my consciousness was separated from my body and working on the inner planes during the hours of physical sleep.

Japikse: It is said you only slept two or three hours a night. Would that be sufficient time to do what you've just been describing?

Tesla: It wasn't always necessary for me to be asleep during this process. I had the capacity to subdivide my consciousness, and a part of me could actually stay on the inner planes, even while I was awake. My form of concentration was unusual. During my moments of intense concentration, part of me could be here on earth and part of me could be on the inner planes—simultaneously. That is not the common phenomenon known as bilocation, by the way. It's something else, and requires a well-developed mind. I did not do this on the astral plane—I did this work in the realm of the mind. The plane of thoughts is much more refined and powerful than the astral plane. It's free of the emotional, sentimental swill most people thrive on, but which must be transcended if you are going to develop clear thought.

You are waiting for me to mention something about my past lives. I am not going to mention much, but I can say that I had earlier lives as a scientist and as a sort of philosopher and teacher. Today the term would be "occultist." During the Atlantean epoch, science was highly developed. Fourth-dimensional principles in physics and engineering were understood to an amazing degree. I participated in those discoveries and applications at that time.

Japikse: I have a question about Atlantis. According to many reports, Atlantis was technologically superior to our present civilization, yet my understanding is that it was a civilization which was primarily focused on the astral or emotional level. Was there a comparable or superior technology then, and if so, how was that technology developed when the emphasis was so emotional and not mental?

Tesla: No problem. Every civilization has its inspired minds. They may be rare, but they are always available. Atlantis was no exception.

Look at the average person today. He leaps into a car or an airplane to go from here to there without pausing once to consider what a technical marvel it is. He thinks nothing of using these technical masterpieces, but would have no idea how to make one himself—or even repair one. There's a tremendous gap between the development of the average person on the one hand and the development of the technology he or she uses every day.

The same gap was common in the Atlantean era. They had great technological achievements, but the average person had no understanding of them. Yet there was one characteristic of the way the average Atlantean viewed technology that I would urgently recommend for today's world. For all their faults, and they were manifold, nonetheless the Atlanteans *respected* intelligence. They respected genius in any line of achievement, be it government, religion, philosophy, art, or science. And the Atlanteans took care of their geniuses. They were an extremely proud race in many ways. They respected people with genius, protected their development, and listened to them. They didn't take popularity polls to see what a good idea was—they listened to their geniuses.

As a result, the best and the brightest of the Atlantean geniuses had a much better chance than today to make a contribution and then build on it. They had a better chance to enrich the work of each other.

The Atlanteans also respected the process of inspiration—especially the psychic side of inspiration. That often made it easier for helpful entities and spirits to poke ideas into the receptive and interested minds of physical scientists and thinkers. Mediumship among scientists was not unknown. Can you conceive, in this day and age, of a college professor being a medium and using his mediumship openly in the conduct of his scientific studies? That could happen in Atlantis. But today, mediums are often thought to be floozy, badly-educated women who bring through trivial comments about what vitamins to take, what dress to wear to a party, or where to take your vacation.

I wish science would realize how valuable the intuitive process is to its work! I know a lot of scientists are studying the few writings and ideas I left behind, to see if they can reconstruct my work. Let me tell them something: the greatest asset in trying to understand what I did would be the development of psychic awareness. Understanding my work certainly requires a basic understanding of physics, mathematics, engineering, and mechanical phenomena, but it requires more than that, too. It requires an awareness of the invisible forces which influence the physical plane. And to understand these forces, you have to have a mind which can reach into the invisible realms of life and deal with the essence of ideas.

It is possible to develop a faculty of clairvoyance and be able to see some of these forces at work. A measure of etheric sight would be helpful in understanding physics or electricity, for example. Some clairvoyants, as you know, are able to see auras and lights around the bodies of humans, animals, and even plants. It's also possible to see the magnetic aura around magnetic or ionized bodies. Psychic experiences such as that would expand the scope of scientific investigation.

But even that is not the type of awareness I'm most interested in promoting. I want to emphasize the idea I mentioned earlier about training the mind to summon the essence of an idea and to synthesize facts and integrate them into meaningful patterns and principles. This means training the mind in intuition and *genuine thinking*.

I'm sure all of you here have had experiences where you have recognized a pattern or meaning in a collection of facts that made no sense at all to anyone else. That is because you have the ability to synthesize ideas. You saw something where others saw nothing. Yet everyone was looking at the same set of facts.

This kind of intuition is a necessary part of all genuine creative work. I don't think science can really progress until this kind of awareness is recognized, honored, and studied. It isn't enough anymore to simply study textbooks—not if you want to produce something worthwhile. You could be an excellent technician or draftsman by just studying the textbooks, but never an inventor or discoverer. That requires an intuitive capacity.

Science is nearing the end of its frontier, and the reason why is because it is so pigheadedly materialistic. Science believes that if you can't push it, knock it, shove it, or measure it, it just doesn't exist. So it ignores most of life. Many scientists have exactly this kind of attitude. They are like the people who used to believe the earth was flat. They need to accept the possibility that the universe goes on beyond their own limited view. Beyond their horizon, there are invisible realms of existence waiting to be explored. That's where science needs to go!

These territories cannot be explored with physical measuring devices, however. They have to be explored with the mind. The new scientist will have to learn to use the mind for *awareness*. It's more than just a tool for collecting and storing facts.

These are not mental skills that the scientist would have to be born with, by the way. They can be acquired in much the same way that mathematical skills can be acquired. They can be studied and learned like any subject.

Gordon: Would you mind talking a while about healing?

Tesla: Well, it's a little out of my field, but I'll try to answer your questions.

Gordon: I was wondering what you could tell us about the energies a healer uses in healing.

Tesla: The goal of any healing is to produce a change in the form or structure of the diseased physical tissues and organs. The end result, therefore, is on the physical plane. But, as I understand it, healers work with many different levels of energy. Some work with physical energy—actually, it is etheric energy. But that requires an enormous amount of physical effort, if the energy is supplied from the resources of the healer. Most healers, I believe, work with astral energies—actually, low astral energies in many cases. Some healers work with higher forms of energy—for example, the more subtle types of astral energy. A few work at still higher levels, on the mental plane. True healers, of course, always work with the soul force. It's even possible to work with energies coming from levels higher than that, although only a few people can do that. I'm not sure that answers your question.

Gordon: Well, is it possible to build an apparatus, using the level of technology we have today, that could do mechanically what spiritual healers do?

Tesla: You could build a device that could do some of what the healers who primarily use etheric energy do. The place to start is with two fields of knowledge already in operation. One is the well-developed science of acupuncture, which deals with the etheric body and the magnetic gridwork. The second is Kirlian photography. In working with these two techniques, you should be able to put together mechanical devices which would be helpful in the healing process.

In point of fact, nerve forces are really etheric currents. These etheric currents could be augmented even by using certain physical apparatuses already in existence. The type of current and the exact site where it should be applied could be determined by studying acupuncture and the evidence which could be obtained with Kirlian photographs. The first type of conditions to examine would be those requiring quantitative changes of the nerve force. If there is too much energy, the object would be to drain it off; where there is a deficiency, you would add new energy. In part, this can be accomplished by opening up pathways of current or by draining off excess energy through alternate pathways. Eventually you would be able to add new energy to the system as a whole.

It would also be possible to work with the different qualities of energy that go into the etheric body and apply them directly to certain points in the system—or to the spinal cord, which is the main trunk for the distribution of etheric energies. I'm talking now in terms of certain low amperage, high-frequency currents, but I can't get too specific. You would have to discover them through experimentation. It might be helpful, for example, to work out the harmonics between specifically-colored lights and electricity, and the use of certain crystals as a type of transformer and focusing agent.

Let me add this thought. If you think of the human body as a sort of biological battery, you can charge it just as you can charge a flashlight battery. If a person's charge were low, recharging the human battery with a weak direct current would have enormous value in revitalizing the system. That would be a major application of what we're talking about. And then, by selected use of acupuncture meridians, you could channel this energy through the body's distribution system and get it where it is most needed. Does that make sense?

Gordon: Yes. Could you also develop an instrument for diagnostics, based on the same principle?

Tesla: I think such instruments have already been developed, if we're thinking of the same thing. To use such a device, it would be important first to study a lot of normal people, to determine the normal configurations, potentials, and frequencies. That needs to be done more thoroughly than it has been.

As you know, I studied the currents which surround this planet as part of my work. I discovered there are certain standing waves and a natural resonance to the whole planet. Based on that discovery, I could then produce many interesting phenomena. In healing, it is important to appreciate the fact that there is a certain resonance that goes with the whole human body. This collective resonance is the sum of the separate resonances of the individual organs and tissues. If healing were approached with this idea in mind, some very delightful things could happen.

Johnson: What do you think of the psionics devices which are being experimented with more and more?

Tesla: There's a great deal of possible application to healing. But there is also an inherent danger in the use of these machines, just as there is an inherent danger in leaving your horoscope around.

These psionic machines—the Hieronymous device and the Del a Warr box—illustrate a principle I worked with all my life. That is the innate resonance which exists between the part and the whole and between the several parts of the whole. A drop of blood in your body is in resonance with the totality of your physical body. A photograph of you is in resonance not only with the totality of your physical body but also a good chunk of your personality and subconscious. And it is possible to apply this principle in many ways.

Just as it is possible to transmit power or radio waves through the air, the same principles can be put to work to transmit healing forces between a drop of your blood and your body. This would have its immediate effect on the etheric body, of course, not the dense physical. The effect would probably be registered subconsciously rather than consciously, but it would still be real. And it would be a manifestation of a natural, harmonic, electrical resonance between the part and the whole.

Johnson: If we had this resonance between the drop of blood and the body, would we also have a resonance with a time yet to come in that person's existence?

Tesla: Yes. Your essence is much less changeable than the dense physical form, and the basic resonance is with your essence rather than the form. Therefore, it would be possible to tap other time periods. It would be easier to tap the past, however, than the far future. The past is fixed, because it has occurred. But the future is variable, within certain bounds. If you tried to use the principle of resonance to tap into the future, you would probably be able to discover the most probable alternative for some course of action, but that would not necessarily be the same as the actual event which ultimately would come to pass.

Johnson: I've been trying to apply this idea of resonance to my painting. I'm not sure what will come of it, but I'm hopeful something will.

Tesla: I am told there is a magical process involved in painting which could have therapeutic effects. I'm also being told to finish my thoughts here. *[Laughter.]*

You do a type of painting which is symbolic of the subject's character. I think it's called an aura chart.

Johnson: Yes. I call it a "psytrait."

Tesla: Well, a good aura chart would have a resonance with the consciousness of the subject you're painting it for. It would stimulate the subject. Does that make sense?

Johnson: Oh yes.

Tesla: I'm really quite delighted you asked this question, because it's a good example of the nature of resonance as I'm trying to describe it. Everything in the physical plane exudes an aura. You don't have to be a human being to have an aura. Physical objects have auras, too. Now, good art should be a form of magic. I would call it a form of esoteric physics. The good artist deals with personal magnetism. In doing your psytraits, you symbolically anchor a portion of the essence of the subject's basic energy system into the form and color on the canvas. You then add to this portrait or aura chart a certain magnetism, a magnetic potential which stays within that psytrait until it is physically destroyed. It would act very much like a radio sending station.

Johnson: I see.

Tesla: As the painting hangs on the wall in the presence of the subject, then, it stimulates something within his consciousness. It could be healing, or inspirational, or just help him get to know himself better.

This is the esoteric basis of effective art. I'm sure you've had the experience of standing before certain paintings—the paintings of real masters of the craft, like Rembrandt, Botticelli, Reubens, or El Greco. As you stood before these masterpieces, they produced a powerful effect on you. Even if you were blindfolded and walked around a museum without being able to see the paintings, you would still recognize the real masterpieces by the strength of their resonance. You would realize that there is a strong, beautiful force pouring off the wall over there. And if you then took off your blindfold, you would discover that on that spot of the wall, there was a rather unusual painting which strikes you with tremendous impact. Good painting is always an electrical art, as well as a psychic art and a magical art.

You might try that—walking around a museum with your eyes shut.

Johnson: I have. I know what you mean.

Tesla: Besides stubbing your toes and bumping your nose, did you pick up anything? *[Laughter.]*

Johnson: I got exactly what you just described.

Japikse: I'd like to ask another question concerning healing. Bob and I have both had the experience of telling people about these interviews and how many good ideas have come through already. And many people have said, "Well, why don't you ask Cayce or Tesla for a cure for cancer?" How would you respond to that? Is the cure for cancer something which can be given out like a recipe for fudge? *[Laughter.]* Or is it more complex than that? And is a cure for cancer necessarily the most important thing we could talk about?

Tesla: The "cure" of any disease would be a very complex issue and, no, it could not be reduced to a simple formula like a fudge recipe. Besides, this was not my work. I would not be capable of delivering the answer even if it could be reduced to a few simple sentences.

For one thing, there are so many different factors involved in a cancer. To really cure it, you would have to address them all. There are environmental influences which have built up over decades of exposure, there are genetic potentials, habits of lifestyle, and psychological patterns of thought and feeling.

There is also the karma of the individual, as well as the karma of the racial group he belongs to. In addition, there are the diseases and indiscretions, physical or psychological, of humanity as a whole.

Japikse: Sounds complex.

Tesla: It is complex. That's why it's so difficult to talk in terms of a real cure for any one disease. Each individual is something like the proverbial drop in an ocean of water. It is not separate from other individuals. If impurities are spilled on one area of that ocean, each drop in that area is going to be contaminated to some degree. It will not be easy for that one drop to become completely cleansed until the whole ocean is cleansed.

I'm trying to describe the actual dimensions of disease. Individuals can be helped individually, but cancer as a whole cannot be cured until the attitudes of humanity as a whole can be effectively changed—attitudes of lifestyle, purpose, habit, and many other things. What might be a useful suggestion for one person with cancer of the colon might not work at all for another person with that same problem because of differences in lifestyle, exposure to environmental toxins, temperament, and karma. However, there are certain ideas that would be worthy of investigation, which would help science deal with the phenomena of disease, and cancer in particular. I've already mentioned appreciating more the role the etheric body plays in regulating health. I've also explained how a lack of vitality in this etheric body can lead to physical diseases. What I might add here is that disturbances in the *quality* of electricity flowing through the etheric body can also be a factor in disease.

Japikse: When you say "quality of electricity," exactly what are you referring to?

Tesla: As I said before, there are many different types of electricity, depending on how you generate it. There can be differences in the sort of cations and anions involved. This can affect both the generation and the transmission of electricity. In the human etheric body, the amount and distribution of certain elements—often, heavy metals—has a great deal to do with the ease of circulation of electricity and the type of current that can flow through the body. The physical body is very prone to be filled up with substances which impair the proper circulation of electricity. These are toxic substances.

A realization of this principle is sometimes approached by students of nutrition. It is not thoroughly understood by them, but they come close to doing the right thing in feeding the body foods that are high in vitality and balancing out the mineral content of the body—and other catalysts—so that the body has a higher quality and intensity of electrical flow and a proper distribution of current.

Quite simply, there is an electrical basis to every life form, and electromagnetic properties. If people are interested in the health of the physical body, they ought to pay attention to the electromagnetic phenomena which are essential to the health of any life form. They ought to investigate the elements of the body that can conduct or retain a magnetic or electrical charge.

Japikse: Would a scientist wishing to do this be most successful if he pursued his investigation along the lines of your earlier comments about investigating the electromagnetic sensitivity of plants, only applying those ideas to humans?

Tesla: That would be one way. For example, he could investigate what augments the electrical charge in a plant. Then he could consider how to alter the charge either to make the plant healthier or to harm it. What is toxic to the plant and what is beneficial? What impairs the electrical field and the flow of electricity through the plant? By experimenting in this way, he might begin to understand what goes on in human bodies. These ideas could then be applied to understanding the way in which electricity in the human body is distributed through nerve channels, and I'm not talking about the gross physical nerves, but the nerve forces of the etheric. There is much to be studied on this subject.

Japikse: I have a question which will seem totally off the subject but will lead back to it. Since you have become a spirit, have you by any chance had a hand in inspiring science fiction?

Tesla: Science fiction is often used to inspire the minds of the masses. As I have stated, my work is with an organization which is dedicated to promoting the proper and full use of the mind and to stimulating people to think. We try to stimulate people to look at life and its manifold phenomena in new and different and more imaginative ways than before. So there are many of us who do use science fiction as a means of sneaking in an idea here and there, as you surmised. It's very useful.

Japikse: I was specifically thinking of a book by Robert Heinlein called *Waldo*. One of the main features of the book is the transmission of electricity from a central power station such as you developed in your lifetime. As the plot thickens, Waldo learns how to draw energy directly from the atmosphere.

Tesla: Yes.

Japikse: That seemed so much along the line of the work you did.

Tesla: Can you imagine that? *[Laughter.]*

Japikse: I thought perhaps you may have had a pinky or two in Heinlein's pudding.

Tesla: Yes, and some things Van Vogt wrote, too. [A.E. Van Vogt, author of *The World of Null-A* and other stories.]

Japikse: Well, here's how this ties in with what we've been talking about. In *Waldo*, when electricity was broadcast from a central source, people began suffering from intense radiation poisoning. Would that actually happen?

Tesla: Oh yes. As you know, microwaves are already being broadcast widely, and there's quite a strong controversy about microwave pollution. People are worried about the harmful effects of living next to microwave towers and operating microwave ovens. There is legitimate cause to be concerned.

Japikse: Good grief! Is my microwave oven dangerous to me?

Tesla: Not really—not if it is properly shielded and you don't stand in front of it ten hours a day. Be sure not to stick your head inside it, though. *[Laughter.]* On the other hand, there *are* grave risks of certain types of radiation pollution. It is possible to get around these risks, however, either by modifying the design and application of the apparatus involved or by putting them in more remote areas.

Japikse: Was this radiation pollution another reason why some of your experimental work never got widely applied?

Tesla: When I was physically alive, I often was driven by the force of a specific idea or its application. I would start with three or four spokes of the wheel and then the wheel would move on and I would suddenly see three or four more spokes. Once I got involved in an idea, many more applications and uses would come to light than I first expected, and I just kept going. It would have been premature to loose many of my developments on society. It might have given civilization indigestion to have to absorb too many marvelous new inventions all at once. *[Laughter.]* So I had some help from time to time from the other side in making some of what I brought to light obscure again. Most of these things were only in my head anyway, so it wasn't hard. When I died, I just took them with me. Who says you can't take it with you? I did! *[Laughter.]*

Here's an interesting point. Sometimes the higher forces of life work very diligently to fertilize the minds of certain personalities with wonderful new ideas or at least the seed of discovery which, when planted, will lead to a new invention, concept, or practice of benefit to humanity. These higher forces are like the parents of civilization. Now obviously, parents do more than just teach their children what to do. They supervise the play activities of the children and protect them from getting into too much mischief. The parents of civilization do much the same. In my work, some of these entities assisted me in making sure some of my ideas were never perfected—or if perfected, then never written down.

I might say the same type of supervision still goes on today, especially in such areas of science as nuclear physics, the space program, and laser technology. Certain possible areas of investigation could not be safely explored at this time, and so scientists are kept from exploring them and diverted into safer areas.

Japikse: Very good. Getting back to microwaves for a

minute, is it possible that there can be some benefits from using them that people ought to be aware of, and not just focus on the potentially harmful factors?

Tesla: Yes. Let me draw an example from nature. Within a balanced ecology, certain plants seem to support other plants by being in the same area with them. In other areas, they might be harmful, however. As the old platitude puts it, "One man's meat is another man's poison." Much the same can be said for electromagnetic radiations. This is why it is so important to study the subtle aspects of these energies.

The closest analogy here is in the phenomenon of sound. Certain sounds, such as rock and roll music, are harmful to life. It has been demonstrated time and time again that rock and roll music kills plants. In contrast, classical music enriches the growth of plants. In the same way, good music helps human beings and bad music harms and confuses humans, and can even make them physically ill. Electromagnetic vibrations can be used and abused in a very similar fashion.

Again, the more scientists investigate and discover how electricity affects the physical system, the more obvious it would be which electromagnetic frequencies are helpful and which are harmful. As I said, a good place to start would be by investigating such things as the bodily composition of trace minerals, the acid base balance, and so on. These are clues that can lead to a better understanding of the electromagnetic human system.

Japikse: Those comments seem to contain a vast number of excellent suggestions. Would it be fair to say that something like microwaves would be harmful to some people but helpful to others? Or would a specific type of wave always be harmful or always helpful?

Tesla: No, I'm afraid ordinary microwaves, broadcast through the atmosphere, are basically destructive. Of course, there are times when you want a destructive force—for example, when you are trying to destroy cancer cells.

Japikse: But the use of microwaves for long distance communication is not really helping humanity?

Tesla: No.

Johnson: I have a question I'd like to ask. How far did you take your work concerning the magnetic field of the planet?

Tesla: Well, that's a pretty far-reaching question, but I suppose I can make a simple statement. We used to think of electricity as some kind of invisible fluid. That's a rather constructive analogy. The earth's magnetic field is somewhat fluid. It certainly is elastic. You can make specific types of waves in this fluid and you can move those waves around. Moreover, just as you can transmit sound and light through water, you can also transmit electricity through the air and through the planet.

Let me give you a different analogy. Where there is air, you can transmit sound; where there is no air, you cannot transmit sound. Where there is an electromagnetic field, you can transmit all kinds of things through it.

Based on these ideas, I think you can jump to a higher octave and see that the sea of electricity which surrounds the planet is both a medium for transmitting power and a source of power. It can be raw power or qualified power. You can transmit information, heat, light, electricity, and many other things. How much farther would you like me to take this?

Johnson: Well, I believe you once stated that you could tap the resonant fields of other planets as well as the earth. There is some suspicion that you were in contact with another planet for fifteen years, and I guess some people have done some rather hysterical speculating about what you were doing during that period. Were you indeed in contact with another planet?

Tesla: Well, we are all in contact with other planets, as you perfectly well know. The energy being transmitted from the sun affects us and, in subtle ways, the energy from other planets does as well. It is perfectly reasonable to contact the energy fields of other planets and apply it.

Let me drop another rock in your pond. An astrologer can be an incredibly accurate diagnostician of the human personality. Correct?

Johnson: Correct.

Tesla: But astrology doesn't have a therapeutic application, does it?

Johnson: No.

Tesla: Well, if you could tap the actual energies of the different planets in this system and work with them, then astrology would have a therapeutic application, wouldn't it?

Johnson: Yes.

Tesla: Do you see what I mean? It can be done!

Johnson: Well, my question was—

Tesla: I don't want you to gloss over the fact that I have just made a *very important* statement. Okay?

Now, was I in contact with flying saucers and little people with three eyes, green skin, and antennae on their foreheads? There really wasn't any need to be in contact with extraterrestrial intelligences. I did not receive any of my inspirations or ideas in that way.

Johnson: I'm not asking for that reason. You made a statement that you had cracked the code for some other planet and apparently were in contact with the intelligences there. I was just wondering if you could clear that up a bit?

Tesla: Well, I will not add to the hysteria of the flying saucer freaks. There are, of course, intelligent beings on other planets, including planets unknown to science. There are several other planets in this system, as well as planets in other solar systems.

Gordon: Are you saying there are planets in this system that have not yet been discovered?

Tesla: Yes. Of course, some of these planets do not have a physical basis.

Johnson: If these planets do not exist physically, where do they exist?

Tesla: A planet can be made of astral matter or mental matter, rather than physical matter. Or it can have an etheric body but no dense physical planet to go with it.

Johnson: Would these be fourth-dimensional planets, then?

Tesla: Yes, although a lot of planets which exist on the fourth dimension also have a dense physical planet with it.

There are also planets where the civilization exists only in the fifth dimension.

Gordon: Can I change the subject?

Tesla: Of course.

Gordon: We seem to be discovering that some of the lower forms of life on this planet are aware of human beings. Is this the case?

Tesla: Of course. Everything has a certain degree of awareness. Sometimes it is pretty dull, but certainly the more evolved animals and plants have a measure of sensitivity which could be called "psychic." They are certainly aware of human thought. We live in the magnetic atmosphere of the planet, after all. And this magnetic atmosphere is filled with human thought. That's the basis of telepathy, and telepathy exists between animals and humans, between animals and animals, and to a degree between plants and plants and plants and humans, as well as between humans and humans.

Gordon: It's accurate to say, therefore, that there is a kinship among all living things?

Tesla: Oh, absolutely. It's an orderly universe. And the first law of creation is that everything must somehow be in harmony with everything else, at least at the level of essence. This is poetically expressed as the kinship with all life. But even the rocks and the soil are alive, although that fact is not widely recognized.

Gordon: Is all of nature capable of mental processes?

Tesla: The higher animals certainly use some mental processes. To a slight degree, some plants do.

Gordon: I guess my question is this: is there an "overmind" made up of all living things?

Tesla: Yes, but it doesn't exist on the physical plane. Just as there is a soul for humans, there is a group soul or inner essence for animals, vegetable species, and even rocks and mountains and streams and clouds. But that is entirely out of my field. These intelligences exist not on the physical or etheric planes, but on higher levels. At the head of all these intelligences, of course, is the creative essence some call God.

Gordon: Could this overmind of nature have a physical influence on man?

Tesla: Of course. It always has and always will. Its activity is not apparent to most, because they don't think about it, but it is active in virtually everyone. It's exceedingly subtle, of course. Anytime a genius develops new ideas or physical forms, be they artistic creations or new machines, to a degree he is expressing the influence of the Overmind.

Gordon: Would that be the purpose that man has been raised for?

Tesla: Well, you can't have creation without purpose. And mankind does serve part of the overall purpose of creation. But I'm not sure I understand your specific question.

Gordon: Is mankind merely the end product of many series of experiments by this Overmind?

Tesla: They are not experiments—they are planned, purposeful stages of evolution. That would be better terminology. What you have now in terms of the human species and intellect is the end stage of millions of years of evolution. It is the latest product—not the finished product, but the latest one.

Gordon: Is contact with extraterrestrial planets one of the purposes of the Overmind?

Tesla: That's not a purpose of the Overmind, no. It does happen, however, because the Overmind of this planet is connected with even more exalted intelligences, such as the Overminds of the sun, the other planets, and a few nearby stars and constellations.

The thumb on Robert's hand exists because it is part of the body. It is not the *purpose* of the thumb to be attached to the body, but being attached certainly helps it fulfill its purpose. Its purpose is to serve the work of the hand, the arm, and the whole person. Is that analogy helpful?

Gordon: I'm not sure. Could the Overmind destroy man if man abuses its privilege here on earth?

Tesla: Well, whatever has been created can obviously be

uncreated. The thumb could be severed from the hand. But such a thought presumes some rather odd premises. First, if the Overmind is smart enough to create man, it certainly knows how to do it right, and there would not be a creation made so imperfectly that it would have to be destroyed. Number two, the concept that God created man and then man got away from Him—that God can no longer control His own Creation—is an absolutely absurd notion. Having manifested His Creation, the Overmind or God or whatever you want to call it continues to maintain contact with it and control it. The control is not total, but it does guide the fundamental issues of life.

The thought that the Overmind would destroy man is not tenable for these two reasons. The Overmind is still in control. That is going to disturb a lot of negative thinkers who are pleased by the thought that the human race is evil and going to the dogs. Such people are simply on the wrong side. They are going to find they have been on the wrong side for a long time. They are not participating in the evolution of humanity. They are mired in stagnation.

Gordon: Still, no other species has damaged the common relationship of nature as much as man. Are we going to live in harmony with the rest of living things?

Tesla: Of course. It is extremely encouraging that man has realized his error in time to do something about it. And, of course, he will. Some people absolutely froth at the mouth about the possibility of mankind destroying one another or the planet. That will not be allowed to happen. There is danger, to be sure, but the danger is being corrected now, and it will be contained before it gets too far out of hand.

Bear in mind that it is through the creative mind of man that evolution proceeds. Without these inventions and technology, which sometimes include byproducts such as pollution, mankind would not progress. But even these problems contribute to evolution. One of the interesting by-products of air and water pollution and the depletion of natural resources is the stimulus which has developed to create alternate energy sources.

This is something I worked on myself. When new sources of energy are discovered, many of the causes of pollution will be effectively abolished. So you see, some of these problems are self-correcting.

Gordon: All we lack in generating new energy sources with the fusion reactor is a better insight into magnetics. Will we get that insight soon?

Tesla: Yes, but not in the way it's expected. The current approach to developing power by fusion will probably be brought to some fruition, but will be so cumbersome it will eventually be abandoned in favor of something vastly more simple. There is a tremendous amount of electricity hitting the earth every day from the sun. This is something that ought to be tapped—and it can be tapped very easily. It would be much more simple and far less of a technical problem in the long run.

Japikse: A question I would like to ask has to do with the fact that both you and Thomas Edison did your tremendous work in electricity at the same time. There is some speculation that you didn't care much for Edison and his work. Would you care to put this into perspective now that you are a spirit?

Tesla: We made our contributions in different ways and worked from different perspectives, to be sure. Perhaps our individual focuses sustained those differences while we were alive. But the important fact about our respective efforts is that we both were productive and managed to contribute to science and humanity. Perhaps I was a bit better in comprehending the theory and fundamental aspects of electricity and magnetism, so I, in comparison, made the more major breakthroughs in basic scientific discovery. Edison, however, was the man who was intensely practical and inventive in developing applications for electricity—for example, the phonograph, light bulb, motion picture projector, and so on. His contribution was more in the area of technical inventiveness. Mine was more in the area of theoretical breakthroughs. But these are generalities. Edison was a brilliant, inspired, and hard-working man.

The bottom line of all this is: is there a practical application? Is something accomplished which will benefit anyone? Too often scientists forget that the practicality of what they do needs to be kept in focus. There is too much foolishness passed off as "basic research"—as if that is license to try to comprehend why some dogs lift their right rear leg to relieve themselves and others lift their left rear leg. *[Laughter.]* That sort of research tells more about the mentality and morality of those scientists and about science in general than it contributes to humanity.

Both Edison and I demonstrated the potential for genius and creativity in humanity. It is well to preach and speculate about the ultimate nature of man, but the only hard evidence that can safely pass inspection is what *people actually do!* Edison demonstrated what it means to solve problems, be practical, and to persist until a solution can be worked out. I demonstrated the virtues of the inquiring mind. The other contributors to this series also demonstrated aspects of the potential for man's genius. These examples are the reliable indicators of the nature of reality—of man and the cosmos—not idle speculations. If more people would look at what *has been done,* they would find the key to what *can be done!*

This scientific principle can be applied to the study of anything—people, universes, society, or even God. I only wish scientists would be more scientific in this way.

Japikse: Is there anything else you would like to say about the direction science should be taking in the future, especially the near future? Is there a definite plan on the part of the group you work with?

Tesla: We are trying very hard to make science more humanistic and scientists more enlightened and responsible. Science is meant to be the tool of humanity, not its master. And the mind is meant to be the tool of the spirit, when it is fully developed. The mind has other uses, of course, but it is meant to be used constructively. It is not an end in itself.

There is always a grave risk when you promote the development of the mind. In some ways, the mind is amoral. It is always possible to use the discoveries of the mind to harm people as well as help them. That's the nature of ideas. Eventually, of course, if you use ideas and discoveries to harm people, other laws are automatically set in motion that bring about your destruction. So, we are not only promoting the development of the mind, but the development of the responsible and moral mind as well.

As I mentioned, we are interested in getting more people to explore the subtle realms of the physical plane. I wish more people would pick up a readily available but not widely-read book written years ago by C.W. Leadbeater and Annie Besant, *Occult Chemistry*. They very carefully described the operation of etheric forces and etheric matter, as seen clairvoyantly. Geoffrey Hodson has done some investigation along those lines, too.

Much greater attention needs to be given to etheric phenomena. There are many applications to medicine, for instance. The fields of spinal manipulation, acupuncture, massage, and nutrition could be greatly expanded by a better understanding of the electromagnetic basis of these applications. Right now the attitude is, "I don't know why it works but it does, so let's use it." *[Laughter.]* That's not thinking as much as it is calculated mimicry.

We are also interested in promoting a greater awareness that, individually and collectively, humanity is part of the electrical and etheric field of the planet. We draw off it to sustain our life, our well-being, our businesses, and our activities. We also pollute it. Not just with physical garbage, but in other ways as well. Your inquiry about microwave pollution is just one element. Man's use of emotions and thoughts is perhaps an even more serious form of pollution.

Japikse: All right. I have one final question. The biography of your life closes with what the author thought was an example of eccentricity but which I suspect was a bit more profound. Apparently you were very fond of pigeons toward the end of your life and fed large numbers of them from the hotel room where you lived. According to the story, you became quite attached to one in particular. When it was dying, it flew into your window and alit in your room. Tremendous beams of light poured forth through the eyes of the dying pigeon. This apparently made quite an impression on you and was a moment of deep psychic and spiritual significance for you.

Tesla: Yes. If you can see them correctly, all life forms glow. All pigeons glow. They glow with the very force of life. They glow with vital magnetism. I had become attracted to pigeons because I was a lonely, old man. This one pigeon responded more to my affection and attentions than the others, and I became very warm and friendly with it. As any life form does, it responded to my love, and we became close.

The report of the incident was a little exaggerated, but now I understand it more fully. Birds of many kinds, but especially pigeons and doves, are evolving in a line which goes into the angelic kingdom, rather than the human kingdom. Occasionally you will find a pigeon which is far closer to its angelic supervision and destiny than others. That particular pigeon was.

As it came toward the end of its life and was about to have its life force withdrawn—to return to its essence or angelic soul—it flew to me, because we had become so close. I sensed the peculiar presence about it and clairvoyantly saw its aura at the moment of death. It was intensely bright. It was that bright because at that very moment the spiritual essence of the pigeon was being reabsorbed by its angelic oversoul.

That experience moved me to make some realizations about the essential spiritual nature of all life and all life forms, and how we all draw our electrical life vitality from the same source. The most visible component of that vitality is, I discovered, light. We, too, have bodies of light which find expression through our more dense, physical form. So do birds. So do spirits. So do angels.

I was awe-struck, profoundly impressed with the true nature of life, the true nature of electricity, and the true nature of light. And that true nature is God.

I would like to end on that note.

BURBANK RETURNS

In an age that rapidly forgets its true heroes and role models in order to chase after the latest sports or movie star, it is hard to imagine that one of the best known and most popular people of the first quarter of this century was a man who grew plants for a living. Luther Burbank was not especially a public man; he seldom left his nursery in Santa Rosa, California. Nor was he a noted lecturer or writer. Nevertheless, Burbank was a household name throughout the world, because even though few people had met or seen him, almost everyone had eaten a Burbank potato, a Burbank plum, or any of dozens of other fruits and vegetables that he had developed.

Luther Burbank had an amazing capacity to produce new varieties of flowers, fruits, grains, and vegetables. Almost by himself, he nurtured the art of plant breeding and transformed it into a modern science that has directly influenced the study of genetics. Over a forty-year period, he developed more than 800 new strains and varieties of plants, including 113 new varieties of plums and prunes, 50 varieties of lilies, and 10 new varieties of berries. In addition to the potato and the plum, he is most famous for his refinement of the rose, the giant Shasta daisy, and the fire poppy.

Burbank's output was so incredible that he became known as the "wizard of Santa Rosa," and many people believed that he had magical powers. To his credit, though, Burbank never claimed to have performed miracles. He was instead a tireless worker who brought a highly developed skill of observation, a deep knowledge of his field, and a capacity to work with great rapidity to his labors. The combination paid off. Like Babe Ruth, who hit so many home runs not just because he was talented but also because he played so much baseball, Burbank produced so many new varieties not only because he was a genius but also because he worked so hard.

But Burbank *was* more than just a hard working plodder. He had a rapport with plants that enabled him to respond to their inner life and design as well as their physical characteristics. Instead of treating the plants he worked with only as objects of scientific investigation, he treated them as living beings. He saw strong parallels between the growth of plants and the growth of children, and worked with a certain conviction that he could direct a plant to grow the way he wanted it to grow— to produce larger and sweeter fruit, to mature at an earlier time, to develop a richer network of roots, to generate new colors or flavors, and so on. The success he achieved is ample testimony that he did indeed know what he was doing.

Burbank related to plants in much the same way that students of the spiritual life approach the nature of man and God—by viewing them as something more than a physical form. Esoteric students learn to direct their attention to the life force and qualities that animate the form of man. In the same way, Burbank directed much of his attention to understanding the inner life of plants, so that he could work more skillfully in improving their visible characteristics. With this perspective, for example, he could conceive of the true being of the potato plant overshadowing a whole field of potatoes. Each potato bush held part of the potential for the multitude of ways the inner design of potatoes was attempting to evolve. And so he knew that if he was discerning in his selection, he would be able to choose those few potatoes that exhibited at least a trace of the characteristics he needed to produce the results he wanted. Once he had made his choice, he then used a variety of breeding methods to quickly regenerate new plants that would exhibit a stronger measure of the desired characteristics, repeating the process until he had achieved his objective.

Left to nature, the same results would have been achieved, but it might have taken thousands of generations to do so. Burbank worked with both speed and brilliance in his use of crossing and grafting techniques, producing many of these innovations in just a year or two.

It is not an exaggeration to say that Burbank was really a mystic in action, a transcendentalist of scientific botany. He was a steward of the plant kingdom, a priest of nurturing growth. And he was a marvelous example of how we can render useful service to promote the evolution of life in the lesser kingdoms.

All of this Burbank accomplished without benefit of a college degree, support of large grants from foundations and universities, or use of scientific laboratory facilities. His education came through his own keen observations. His financial support came by selling the rights to the new varieties he produced. His laboratory was nature itself—and his brilliant mind.

Luther Burbank was born in Lancaster, Massachusetts in 1849. At the age of 21, he purchased a farm near Lunenberg, Massachusetts. Within a very short time, he had produced the Burbank potato, which remained one of his most famous and useful varieties, greatly increasing the food yield from a crop of potatoes. With the money he realized from the sale of the rights to the potato, he traveled to California and reestablished his nursery in Santa Rosa, where he worked for the rest of his career, until his death in 1926.

To me, Luther Burbank is a brilliant example of the highest kind of Christian stewardship. Although not an overtly religious man, he had a deep respect and love for divine life as it manifested through all life forms—plants, animals, and humans. He was a scientist by profession, but more a mystic by persuasion. He approached his work as though he was thoroughly connected with the inner life of spirit, both his own and the inner life of the plant kingdom. Truly, he was a man who was in harmony with life.

This kind of quiet but effective service is often forgotten in the constant uproar and pleading on behalf of the poor and the downtrodden. Humanity has fallen into the bad habit of thinking that service is something done only for the disadvantaged and the distressed. It is true that the poor and the ignorant need whatever help we can give them, but it is unfortunate that the concept of service and stewardship has been so exclusively linked with the desperately needy. *Anything* which contributes to the growth of consciousness and the unfoldment of divine potential must be regarded as service. And the highest service of all, whether it is offered to humans, plants, animals, society, or any other element of life, is the service that helps some part of divine life *go beyond the natural limits of its potential.*

This is what Burbank demonstrated so admirably. He served as an agent for the inner life of the plant kingdom, helping develop new and better physical expressions for *hundreds* of species. In many cases, in fact, he did what the plants themselves could not have done, thereby greatly enhancing their development. All human service should be similarly motivated—to accelerate the natural evolution of whatever aspect of divine life is being served.

Burbank also gave us a wonderful example of the power of nurturing and healing love in action. It was the love he had for plants that gave him the strong intuitive rapport with their inner essence, and enabled him to draw forth a full measure of their potential. He demonstrated the kind of nurturing love which should be the dominant quality of child rearing, medical practice and nursing, and all social services. The fact that Burbank was a scientist only serves to make his example even more noteworthy.

The idea of a scientist "loving" his work and his experiments may seem odd, but it produced one of the most prolific geniuses of our century. Every scientist can benefit from studying Burbank's life and his holistic approach to work, as it teaches us that the proper study of life is the phenomena of life itself. Burbank regarded nature as the one great textbook, and sought to be guided by its rules only, not the rules of human "experts" and "authorities." He likewise regarded evolution as the one great experiment, an experiment being conducted not by scientists in white coats but by nature herself. And it was his great joy to participate in this one great experiment. To me, that is the mark of the true scientist—in this case, a scientist for all times.

In the interview that follows, Burbank speaks at length about the inner life of the plant kingdom, how it is evolving, and what we can do to cooperate with it. It was quite evident to me as I chatted with him that his knowledge of these matters was not in any way a superficial one. His comments and observations ring with the depth and wisdom of someone who is working intimately with archetypal forces. Indeed, during one of the breaks in taping the interview, the medium, Paul Winters, commented that when I asked a question about the evolution of plant life, an incredible vision of the life cycle of the plant—not just one plant but generations of plants—flashed into his awareness. It was obvious to him that Burbank was very much at ease handling this kind of comprehensive vision of life, but it was almost overwhelming to Paul. It was apparently a fourth-dimensional representation of the answer—which was translated only inadequately into words.

This is always one of the limitations of mediumship, even the high level of mediumship we try to present in this series. It is very difficult to clothe in words the full realizations that these geniuses, working with archetypal forces, are able to tap. It is hoped, however, that enough of the power of these ideas does come through that they will inspire those who read them to contact the archetypal insights themselves.

Burbank also comments quite eloquently on the nature of

love, the work of stewardship, and the divine force of growth. In fact, he presents one of the most complete statements about growth that I have seen anywhere.

I am joined in asking questions by my friend and colleague, Carl Japikse.

Burbank: It is wonderful to be here. How does this work? Do I get an opening statement?
Leichtman: Absolutely.
Burbank: Good, because I would like to begin by commenting on how I viewed my work and where it fit into the whole of creation. One of the major aspects of this planet which is not correctly understood by mankind is the plant kingdom and what it represents. In my life as Luther Burbank, I worked to demonstrate, at least to some degree, that plant life is part of the whole of life; it is a physical manifestation of something divine. And it can be dealt with in much the same ways as the other aspects or expressions of divine life. The outer form may be only plants, but the inner life is divine.

Is this making sense to you?
Leichtman: Yes, it's a very cosmic statement. Do you think you achieved that goal?
Burbank: I'm not sure. Many people misinterpreted what I did, in that they viewed it in terms of my work with specific plants, rather than my work with the plant kingdom as a whole. I had not really anticipated that the public perception of what I was doing would be so different from what I actually was doing. But this is what happened. I am known far more for the new varieties of plant life I developed than I am for what I revealed about the plant kingdom and its divine origins. My work is studied more from the angle of reproducing my skills in developing new varieties of plants than from the perspective of my understanding of the divine life of plants.
Leichtman: We have become more absorbed in the details of your work than in the ideals you were serving and the archetypal qualities you worked with.

Burbank: Yes. And in a sense, the statement I was trying to make was not made. In fact, whenever I tried to put my work in the correct perspective, I was looked upon as slightly crazy.

Japikse: Would it be fair to say the statement was made but not listened to?

Burbank: Yes.

Leichtman: Yes, your life work certainly made this statement—and far more profoundly than any words you might have used to describe it. And I would think that any perceptive person reading one of the biographies of your life would sense the deeper meaning of your work, at least to some degree.

Japikse: Even if the biographer himself didn't.

Burbank: I hope so.

Leichtman: Well, as I was reading about your life, it was abundantly obvious to me that you liked to work in relative isolation—that when other people were around, they interfered with your perception and the quality of your work. And the inference I drew from this was that you did your work in much the same way as angels and elementals would work. This simple fact, which to most people might seem antisocial, suggested to me a much larger context and meaning in your work.

Burbank: That is quite perceptive. The ability to understand plant life and work with the energies which nurture and support it does entail a special receptivity to the plant kingdom as a whole and the inner life of the specific plants you are working with. And this special receptivity is very definitely affected by chatter from people who are not attuned in the same way. It was very easy for me to establish this kind of rapport with plants, but it was also very annoying to have the rapport interrupted—especially if I had been in this state of rapport for any significant amount of time.

This was a chronic problem, but it was something I learned to deal with.

Leichtman: When in your life as Luther Burbank did you become aware that you had an unusual receptivity to plant life?

Burbank: At a very young age, before I moved West. It

developed very gradually. I began to talk to plants as a child, and it was clear to me that they were listening. And I was able to understand what they were experiencing—not verbally, in clear-cut, precise words, but intuitively.

Leichtman: Yes, I can recall in my own childhood being aware of the moods of plants, knowing if a plant was happy or unhappy. For instance, I could tell they were delighted after it had rained and they had all had a drink. *[Laughter.]*

Burbank: Yes. Many people look at the care of plants as a mechanistic process, or like ordering off a menu. Every Tuesday at 9 the plant must be given 4 ounces of water and every third Wednesday you must add three grains of fertilizer to the soil. And they stick militantly to the schedule. But this isn't the case at all. Just as humans need more or less food and more or less water depending upon the weather, their activity, and many other factors influencing them at the time, it is the same with plants. They should be fed according to their needs. As you well know, many people gain excess weight from eating by the clock rather than when they are hungry. And the same is true with plants. They need water and fertilizer, but their health can suffer if they are force fed when they don't need to be.

Leichtman: And each plant has its own natural cycle, I suppose.

Burbank: This is true, but even cycles can be affected by changing needs. Nothing replaces the basic link of communication between the human and the plant—which is *love*. Feeding and fertilizing a plant can be a mechanistic process, but if you are really going to *care* for plants, you must be able to treat them with genuine and warmly felt love.

Leichtman: Of course, some people who love their plants—or even their children, for that matter—do not listen to them or observe their real needs.

Burbank: That's not a mature expression of love. It is love without discrimination. These are the kind of people who love their plants so much they water them every day, and end up drowning them. Love should not be confused with overindulgence. Loving a plant—or a child—means understanding it and caring compassionately for its true needs.

Leichtman: This seems to imply, then, that you have to treat plants as a life form with intelligence and feelings and some capacity for self-expression, not as a well-pampered *objet d' art.*

Burbank: That's right. Pampering is not to be confused with love.

Leichtman: What if a reader wants to develop the kind of sensitivity you're talking about? Is there some way he or she could use his imagination to identify with the plant—perhaps think of it as a little child or something?

Burbank: I took the attitude that plants are my equal—that they have a divine life within them just as I do. And while I am a human and part of a different kingdom than plants, plants have the same right to be here as I do. I also knew that if a plant had not been under my care, and had been in its original habitat in nature, it would have been well taken care of. So my link of communication was to duplicate as much as possible the same loving, caring relationship with plants that the divine life within them would have provided if they had not been under my care.

I guess I could be accused of playing God.

Leichtman: It sounds to me like a parent/child relationship.

Burbank: Yes. I tried to provide for the life of the plant with the same parental understanding that God would provide in the natural habitat.

Leichtman: That sounds like a very good description of a strong nurturing attitude—not quite equal, but nurturing.

Burbank: Well, it was equal in that my goal was to provide an environment in which the plant could live by itself. The plant was not dependent on me as a young child would be dependent on his parents. But it was dependent on me to provide the basic environment so it could live and become strong.

Leichtman: I see. You are really giving us an outstanding example of stewardship then—an example we can all learn from, even in the way we deal with one another, as well as plants and all of life.

Burbank: Yes.

Leichtman: Incidentally, I can't help but notice . Do you see how attentively these plants are listening? *[Laughter.]*

Burbank: Of course. They are all friends of mine.

Leichtman: They really are. They are leaning forward and hanging on every word. *[More laughter.]*

Now I've lost my place. Let's see. Oh, yes. It's not uncommon for compassionate psychologists and counselors to do something similar to what you are describing when they relate to their patients. They may even sense, as they are talking with the client, that they are also in rapport with his subconscious or supraconscious. Is that what you did with plants? Did you try to relate to their "higher conscious self" or their angelic essence or whatever it might be called?

Burbank: I didn't do that consciously, but the language of love, when genuinely expressed, automatically sets up this kind of rapport. In retrospect, yes, I think I did have an understanding of the Being Who expresses Himself through the plant kingdom. Now, this is a concept that is not easy to grasp. And I did not know, during my life as Luther Burbank, that this Being even existed, although I assumed that there was such a Being.

I'm being nudged here. We may be getting too esoteric.

Leichtman: Well, in the commentaries I've read about your work, they all referred to your uncanny ability to perceive the ideal for a particular plant—that you could walk down a row of flowers and pick out the exact characteristics which would let you create the effect you sought. There might be a thousand poppies growing in one section, but you could pick out the one that had just enough of a trace of red to produce a perfectly crimson poppy. So you seemed to be able not just to recognize potential but also recognize an archetypal essence or divine pattern that could be refined and strengthened generation after generation.

Burbank: In the same way that your course deals with the relationship of the personality and the inner being, my work was to strengthen and take advantage of the relationship between the "personality" of a plant and its inner being.
Leichtman: Yes.
Burbank: And the same problems that people encounter I encountered as well, but with a markedly different set of rules. The two of you are able to attune yourself to the inner being of a person and look at his flaws from the perspective of his ideal design for growing. You can see where he is expressing the divine qualities of the soul, and where he is not. And you can help him see this, too. I did the same thing with plants. I tuned into the inner essence of a plant and examined how well its physical form was expressing its ideal design. I could see what its flaws and its strengths were. And that let me take steps to encourage the growth of the strengths and eliminate the flaws.

Now, the big difference between working with humans and working with plants is this. You can see these flaws and strengths in an individual, but you can't take steps to actually change them. All you can do is report what you find and try to encourage the person to initiate these changes himself. But I could effect the changes I wanted in the plants, because I was able to supply the outer conditions the plants were lacking. The plants were not suffering from psychological problems or resistance to growth. It was just a case that something was missing in the outer conditions of the plant's life that should have been there. I tried to supply it.

Leichtman: The seed was okay but some other factor was missing that impaired the fullest expression of the plant's divine potential.

Burbank: Yes. And it is also important to observe that the relation between conditions when a plant is young and its capacity for perfection when in full flower is very much the same as the relation between a person's environment when young and his behavior as an adult. If conditions are not proper when a plant is first sprouting, this will affect that plant's growth throughout its whole existence. So in a sense, if we can directly examine the relationship between the inner life of a plant and its outer conditions, we will be able to appreciate, in effect, the whole of its life history.

Japikse: Can you give examples of how the condition of the sprouting plant affects the mature plant—perhaps an example of improper conditions and another of proper ones?

Burbank: Sure. In fact, I'll give you two sets of examples. Let's start with two seeds from the same parent plant—I'm using the word "parent" for want of a better term. These two seeds are the product of the inherent qualities and health of their parent. Now, if you take those two seeds and plant them in exactly the same conditions, the chances are very good they will both develop into equally healthy plants, because they both have approximately the same potential for health. So, in this case, to give you an example of healthy or unhealthy conditions, we would have to plant the seeds in different soil, or care for them differently as they are sprouting. These would be the factors that would produce different results when the plant reached full flower.

But then we can also look at two seeds from different parent plants, one being healthy and the other being unhealthy. If these seeds were planted in identical conditions, they would still grow up differently, because they have different innate potentials. So selection of seeds from healthy, mature plants is another way to produce desired results in the next generation of plants. This depends on being able intuitively to discern the potential of the parent plant to produce the effects you want in the successive generation.

Does this make sense?

Leichtman: Absolutely. And to back this up, it is said that parents who conceive a child while drunk often end up producing a child who has all kinds of psychological problems. It is as though a bad seed is invoked and the conditions of development are then impaired. That seems to be a human counterpart to what you've been describing.

Burbank: Yes.

Leichtman: So your comments about the plant kingdom seem to be pointing to certain principles of development which apply as well to the human race, even though their application is different.

Burbank: Of course. And these principles are vitally important to the other kingdoms as well. One other point I should make in this context is this: the physical expression of a plant could be less than perfect but still be optimal for that particular plant.

Leichtman: That sounds like the way we define human health—that health lies in the capacity of a person to use his body in mature, productive, and healthy ways. And this means that a healthy person does not always have what is conventionally thought to be a perfectly healthy body.

Burbank: Yes, it's more a question of how well the outer form is expressing the inner design. Some people, as you know, believe that bigger is better, but that is not always true in plants. I understand strawberries are being bred larger and larger nowadays, but it's usually the very small ones that have the best and sweetest taste. They are just breeding out the taste.

Leichtman: Yes, they are doing something like that with the tomato, too, I understand. *[Laughter.]*

Burbank: Another interesting side note is that the physical health of a plant has very much the same kind of persistence as psychological habits in humans. And so, if you want to help a plant become healthier, or more expressive of its divine potential, you have to be able to understand and appreciate this momentum of health—and that in fact you are dealing with a momentum and not just a static set of physical conditions.

Leichtman: I'm not sure I understand. You're talking about momentum, but what is moving? It's nothing physical, is it? Is it the life force of the plant? Or something that is not apparent physically?

Burbank: I guess I would call it the "subconscious" of the plant.

Leichtman: If you were to put that into our fancy occult labels, are you referring to the astral body or the etheric body of the plant?

Burbank: It's in the etheric body of the plant. I'm talking about the momentum of habit, not actual movement that could be seen physically.

Japikse: The plant doesn't uproot itself and move four feet away, where the soil is better. *[Laughter.]*

Burbank: No. Now, one of the techniques I used to change this momentum was to put the plant through what amounted to an extended fast. I didn't cut off its access to sunlight, but I did restrict its supply of water and fertilizer for a period of time, so that it was, in effect, on a fast. My intent was to make the plant use up all of the stored energy which feeds this momentum.

Leichtman: Human fasts are used this way, too, I believe. It is known that people undergo psychological changes after fasting for a week or two, and some doctors are using fasts quite effectively to help produce a climate in which enlightened change can occur.

Burbank: Yes. Now, you must keep in contact with the plant while it is on its fast, to make sure you are not killing it. But when you bring it off the fast, you can then start to redirect the health of the plant by being very selective in the nutrients you give it. And what you will find is that the growth that occurs after the fast expresses much different characteristics than the growth that occurred before the fast. And so you end up with a plant that is remarkably different than it was.

Leichtman: How much do the attitudes and intentions of the person caring for the plant affect this process?

Burbank: The attitude is absolutely critical. If you withheld water and nutrients as a punishment, the plant would not respond.

Leichtman: It would be like a parent telling a child, "Behave, or I won't feed you." *[Laughter.]*

Burbank: On the other hand, if the fast is undertaken with the attitude that "this is good for your health and I love you and want you to become a better, healthier plant," then there can be remarkable results. But it must be undertaken in the spirit of love, of always taking precautions to assure that you would never let the fast go to the point where the plant would become unhealthy.

Leichtman: How does this affect the genetic limitations of the plant, though? Obviously, you can't take a poppy and turn it into a potato.

Burbank: Of course not.

Leichtman: Well, how far can you go? Just what is the latitude? You sound as though we could start with a single species and come up with a wild variety of different types.

Burbank: It is possible, but the design of the inner life of the plant would limit those changes to variations within the species.

Japikse (picking up a chocolate chip cookie): You've been talking about fasts; are there other ways to motivate plants, too? For example, is there an equivalent for plants of chocolate chip cookies? *[Laughter.]*

Leichtman: It's probably some kind of fish-based fertilizer. *[More laughter.]*

Burbank: The equivalent of a chocolate chip cookie for a plant would be a moment of intense, devotional love from a human.

Japikse: Ahh.

Leichtman: Plants seem to be more responsive to the power of love than animals or humans, or am I misreading that?

Burbank: It's not that they are more responsive; it's just more a part of their total being.

Leichtman: When I tune in psychically to flowers, particularly to roses, they are just bursting with joy. They are very much aware that they are beautiful and smell good, and they love it. They are almost smug. *[Laughter.]*

Burbank: Yes.

Leichtman: Well, since we've wandered into this territory, perhaps now would be a good time to discuss the role the plant kingdom plays in life.

Burbank: By and large it is fairly obvious, but I suppose we can restate it for the purpose of this interview. We would have to start by saying that the plant kingdom provides the basis for sustaining life on this planet. It provides the oxygen the rest of the life on the planet needs to exist, and also much of the food which sustains life. It provides for the physical needs of the planet.

Leichtman: What would the esoteric role of the plant kingdom be?

Burbank: Part of it is to teach humanity the proper use of emotional responsiveness. The Being Who is the plant kingdom is basically involved in developing responsiveness on the upper astral plane. I am being told that this is the right terminology.

Leichtman: It is now. *[Laughter.]*

Yes, I would agree. I don't know of any plants that respond favorably to hate or fear or lust, as certain humans and animals do. In fact, it has been amply demonstrated that anger, fear, jealousy, and rage are lethal to plants. House plants commonly wither when there is too much tension in the home.

Burbank: One of the keys to right emotional responsiveness is sacrifice. And the plant kingdom demonstrates this principle on a daily basis. It sacrifices itself in love for the animal and human potential of this planet. How many other kingdoms of life on this planet are developed basically to be sacrificed—to be killed and eaten for sustenance?

Leichtman: None willingly.

Japikse: That is certainly true for crops. What about ornamental flowers and trees?

Burbank: They are an expression of the beauty of God, and that teaches us another key lesson about the higher levels of emotional responsiveness.

Leichtman: It is nourishment of a different type.

Burbank: Flowers are usually an integral part of important and meaningful human ceremonies and celebrations.

Leichtman: Or should be.

Burbank: Many of them also have healing qualities. And trees provide firewood for warmth and cooking and wood for shelter.

Leichtman: They also seem to sometimes have the ability to stimulate inspiration.

Burbank: The best way to explore the esoteric value of the plant kingdom is to look at all the contributions plants make physically, and then try to assess the inner dimensions of these contributions. I don't think we want to take the time here to do this, but it would be an interesting exercise. I hope some of our readers will take this as a hint and explore it.

Japikse: Can you elaborate on the plant kingdom being an entity evolving primarily in the higher astral plane? Some people might find it difficult to imagine a plant having an emotional expression, because they equate emotions with a surge of adrenalin or getting angry or mad. They think that you can't even have emotions without a human nervous system.

Burbank: The different kingdoms of the planet evolve at different levels of matter and energy. The evolution of the plant kingdom occurs through the accumulation and expression of positive emotional qualities—devotion, love, beauty, joy, and so on. Humans evolve through the use of emotions, too, but the growth of humans emotionally lies in acquiring *mastery* of the emotional nature, as opposed to simply accumulating emotional qualities.

Japikse: Can you give us any idea of how much evolution has occurred in the plant kingdom? This is obviously something human history hasn't recorded. What has been the accumulation of positive emotional qualities by the plant kingdom over the past million years or so?

Burbank: On a scale of 0 to 10, I would say the plant kingdom has reached a level of 8.5 on its evolutionary scale.

Leichtman: Do we have more flowers now than a million years ago?

Burbank: Oh, absolutely. The crowning evolutionary

achievement of the plant kingdom is expressed through the flower. Most weeds, by contrast, are very low on the evolutionary scale. Of course, as the evolution of the kingdom progresses, even weeds become more and more beautiful.

Leichtman: Are you sure you want to use the term "weed"? Some weeds are very good plants; they are only weeds because humans don't want them in their yards. A dandelion is a perfectly beautiful flower; crab grass is not. But they are both considered weeds by most suburbanites.

Burbank: That's right. Weeds are sometimes in the eyes of the beholder.

Japikse: Will peas taste better in the final stage of evolution? Do they taste better than they did a million years ago—or were there even peas a million years ago?

Burbank: The evolution we are talking about here is not so much the development of the qualities of a single species as it is the unfoldment of the roles played by the plant kingdom. And as the plant kingdom evolves, it becomes more and more useful. Food is one role the plant kingdom has developed, shelter is another, ornamentation is a third, and so on.

You know, the real question may not be how much the plant kingdom has evolved, but rather how much humanity has evolved in learning to use the plant kingdom wisely and to its fullest potential. The human race has been enormously blessed and enriched by the presence of the plant kingdom on earth. And we have learned to use it in many useful ways. But the human race has a long way to go before we really tap the full potential of plants on this planet.

In times before we had developed much civilization, plants frequently provided us with the only real touch of beauty. We didn't have the magnificent creations of Botticelli, Beethoven, Bach, or Rembrandt to inspire us. The beauty and perfume of the plant kingdom was one of the few sources of inspiration, one of the few tangible parts of life that suggested that there may also be intangible, invisible qualities and realms which are important, too.

There is much we can learn still from the beauty and the perfume of plants. Of course, plants also play a significant role in healing—not just physical healing but also healing for the emotions.

Leichtman: What more can you say about the use of plants and flowers for healing? Is there a branch of the plant kingdom evolving especially to supply these needs of the human race?

Burbank: Yes, there is. But again, it is not a question that humans have to wait for the plant kingdom to hurry up and evolve so plants can heal us. *[Laughter.]* The potential of plants to be healing influences, both physically and emotionally, is already developed. Humans have to learn how to take advantage of this potential.

Let me say this. While many herbs are used in the treatment of physical conditions, their most potent effect is on the emotional body, which then can facilitate the development of new and healthier patterns in the physical body. The medicinal qualities of herbs and flowers relate more to the emotional nature of humanity than the physical. And so the area where humans should focus their attention in learning to tap the healing potential of the plant kingdom should be exploring the interaction at the emotional level.

Leichtman: In light of those comments, do you consider the work of Dr. Bach a modern breakthrough in the use of flowers? [Dr. Edward Bach was an English physician in the first third of the century who successfully experimented with the use of certain flowers in the treatment of specific emotional problems. His discoveries are described in his book, *The Bach Flower Remedies.*]

Burbank: Most definitely. And there is much more work to be done. Dr. Bach's work provides a major starting point for tapping more of this potential.

Leichtman: Dr. Bach used plants somewhat differently than most people do. Plants are usually used to treat physical conditions, but he specifically used them for emotional conditions.

Burbank: And this is closer to the truth.

Let me say this. As I have indicated, plants in general and flowers in particular are accumulators of vitality and high-quality emotional substance. By understanding and appreciating this feature of the plant kingdom, and learning to tap this resource, humans could do a great deal to heal and enrich their emotions—and, by extension, the physical body. They could heal the negativity of their emotional reactiveness and enrich their moods and attitudes with the love, joy, peace, devotion, dignity, and affection which the plant kingdom has been accumulating throughout the whole of evolution. These qualities are inherent properties of plants!

Most of the use of herbs has focused on the physical benefits of plant extracts as medicine for the body, but this is only one facet of the healing power of plants. The real power lies in healing the emotions, and this has been barely tapped.

Of course, before humans will be able to tap more of this healing power of plants, they are going to have to learn more about their emotions. This doesn't mean that people need to learn to be more emotional! The full range of emotional equipment has been pretty well developed. What needs to be developed is a greater appreciation for the differing qualities of emotions and the capacity of the human to refine his or her use of the emotions.

The problem is that there is an amazing lack of interest in doing this. People want to be happy but they don't really want to improve themselves. A lot of people are actually delighted to be mean and petty and vicious and angry when it seems to serve their purpose. They aren't the least bit contrite about using ugly, lower emotions to get their way. But these ugly, lower emotions are repulsive to the plant kingdom, and that makes it difficult for the individual who indulges in them to be very responsive to the healing potential of plants.

Leichtman: Well, that comment suggests another question. Where do bacteria fit into the scheme of things? I'm referring to the fact that bacteria are primitive plants, and sometimes they are beneficial to man—they refine our garbage, for instance—but sometimes they are rather deadly. Pneumonia, after all, is often a bacterial disease.

Burbank: I think it is fair to say that it is not a problem of the plant kingdom but rather a problem of the human kingdom. The mere existence of bacteria—or even poison ivy, for that matter—is not what causes the problem for humans. The bacteria themselves are not inherently dangerous. They only become dangerous when certain conditions exist within the human system.

Leichtman: You are saying, then, that it is our responsibility to keep our bodies in such a condition that disease-causing bacteria cannot make us sick?

Burbank: I am saying that the capacity of a virus or bacteria to cause harm is directly proportional to the health of the human being—not the severity of the virus or bacteria.

Leichtman: So that even though a disease may involve a bacterial inflammation, the true cause of the disease is not the bacteria but some other diseased condition, such as a weakened or malnourished physical body.

Burbank: Yes.

Leichtman: Okay, that's clear. I would like to leave that and ask something else about the varieties of plants. I have often noticed on my sojourns in the astral plane that there are a wide variety of incredibly beautiful plants that do not exist in the physical plane as far as I know. Are these some future century's models for plant life on earth, or what?

Burbank: Let me answer this question by drawing an analogy with the human kingdom. There are many human souls who at present are not taking part in the physical life and activity of this planet because conditions are not conducive to their appearance. And there are likewise many species of plant life that are not currently part of the physical life of the planet, because it's not beneficial to their development at this time.

Some of these plants which presently exist only at the inner levels could be considered the "adepts" of the plant kingdom, if I may use that term in this context. As the planet evolves, and proper forms for their manifestation develop, they will begin to appear. Others, of course, would be plant forms that have been on earth but have become extinct. The disappearance of a species of plant life on earth heralds the end of the necessity for that form in the evolution of the plant kingdom.

Leichtman: Yes, I notice that occurring in the animal kingdom. Many of the predators are dying out. And if we can beat off certain people in the government who want to protect rattlesnakes and grizzly bears, they may be able to make their exits on schedule, too. *[Laughter.]*

Japikse: Which—the grizzly bears or the people in government? *[More laughter.]*

Leichtman: Well, both are predators that have outlived their usefulness. *[Guffawing.]*

Japikse: Are you saying, then, that these astral flowers are not just on the drawing board—they are actually alive and growing and prospering at the inner dimensions?

Burbank: Of course.

Japikse: They are not just a static idea in God's imagination?

Burbank: No.

Leichtman: Can you say more about what conditions have to change before they can manifest here on earth?

Burbank: There are two basic factors to consider. First, these more evolved plants need a stronger emotional environment of love, compassion, and warmth than is currently available in order to appear on earth. So the emotional climate of the planet as a whole would have to improve. I am speaking of the emotional environment created at present by humans. It is not suitable for these more advanced plants.

Second, it must be understood that no life form appears on earth magically, suddenly materializing out of nothingness, even if the conditions are ripe. These plant adepts would need physical vehicles through which to manifest. And these physical vehicles would have to be provided through the creative work of enlightened humans, working with the principles of cross breeding and seed development to create more advanced forms for the expressions of plant life.

Leichtman: Are you talking about genetic mutations?

Burbank: Yes. To pull together several of the themes we've been talking about, I am talking about the plant kingdom taking the next great step forward. Let me go back to my evolutionary scale of 0 to 10.

Japikse: The Burbank Scale of Growth. *[Laughter.]*

Burbank: Sure. Initially, all kingdoms of life are given the potential and the conditions which will take them up to point 10 on the scale. Then, to go beyond that point, they must make a quantum leap forward. Moving up to 10 is merely the work of fulfilling the preordained potential of that kingdom, but going beyond 10 requires a conscious effort to assume responsibility for further evolution. Humans individually come to this point as they evolve; the human race as a whole will eventually have to face this challenge and meet it. It is part of the design of evolution. But because part of that design of humanity is to be stewards of life on earth, we must recognize that we also have a responsibility to aid the plant kingdom in going beyond 10. Plants have a marvelous capacity to reach their full natural potential on their own. But they do not have the capacity to consciously decide to go beyond this natural potential. And so they are dependent on the human kingdom to provide the conditions necessary to continue to evolve once they have reached the completion of their natural growth.

Leichtman: Of course, the plant kingdom seems to be uniquely able to inspire humans to provide this assistance.

Burbank: Of course.

Leichtman: I refer to people who breed roses to produce a better rose or breed corn to produce a tastier ear of corn. Frequently, they almost seem *drawn* to this work.

Burbank: This is true. Humans have a unique responsibility to provide increasingly evolved forms for the expression of plant life on earth. The plant kingdom has evolved on its own and will continue to do so in the future, but *major* evolutionary changes will have to come through creative human beings.

Leichtman: I take it there is a payoff for the human kingdom.

Burbank: There certainly was for me. The people who learn the most and grow the most rapidly are those who are helping others learn and grow. Anyone who is a parent ought to know this. The infant must depend on his or her parents in order to learn the basic lessons of growing up. But the parents, if they are at all intelligent, learn far more from the child than the child learns from the parent. The parents learn to be compassionate, nurturing, and kind; they learn to make sacrifices and fulfill responsibilities. These are far more advanced and important lessons than the child is learning. The same idea applies to learning to help the plant kingdom evolve.

Japikse: Perhaps it would be interesting if you would comment on what you learned through the work you did as Luther Burbank.

Burbank: I guess I would say that one of the main things was that you can learn an awful lot more from life itself than from reading books or listening to the opinions of experts. In my case, I found that the plants I worked with were the best teachers I had, but this would naturally vary from individual to individual, depending on his or her work and interests.

My point is this. Too many of the "experts" feel that their expertise gives them the authority to make up the rules. Their learned opinions become rules to them, not guidelines, and this traps them. Their work becomes limited, because they start to think of themselves as the authority, instead of life. You have to learn to put this kind of arrogance aside and see what the actual phenomena of life have to teach you. And if you do, you will find that life can teach you far more than scientific texts can. The assumptions of science about plant life are often wrong, just as the assumptions of psychologists about human nature are often wrong and the assumptions of physicians about health are often wrong.

Now, please do not misunderstand me. I am not advocating wholesale intellectual paranoia. What I am suggesting is that the primary source of information about life ought to be all of the phenomena of life. Life itself should be your primary teacher.

The second thing I learned is that you've got to get busy and make a contribution to life. You don't get anywhere by sitting around speculating and theorizing all the time. Achievement comes by getting involved in your work and being willing to put forth a great deal of effort for long periods of time. A lot of people simply do not have the courage or perseverance to sustain activity long enough to become successful. They don't realize how much hard work is involved in being successful; they expect it to come overnight, and when it doesn't, they become discouraged and give up. The reason why I was so successful was that I worked harder than 10 average people combined. I would grow thousands of plants for a single experiment, and out of those thousands I would always have a few that produced the results I sought. Other people might only work with 50 plants in an experiment, so their success was minimal. You have to work hard, use your imagination, experiment, and take a few risks. You can't take a handful of seeds to a laboratory and raise two of them in a five percent carbon dioxide mixture, and another two with no carbon dioxide, and two more in some other conditions, and expect to get meaningful results. That is trivial research and wastes a lot of time. The goal of scientific work should not be to produce a scientific paper, but to produce something worthwhile.

I knew I was producing something worthwhile for humanity, whether it was a better variety of plum or a more beautiful dahlia or a tastier strawberry. I knew my work would be appreciated by people throughout the world, and that was an important motivating factor. I also knew with absolute certainty that nature was on my side. I knew that nature was evolving in these directions and that I was assisting nature. I had found my niche in life and was able to work with a high degree of harmony with nature. I sensed that I was participating in a great experiment, which is really the evolution of life on earth. If someone had asked me what I was doing, of course, I would not have been able to articulate this idea. But I sensed it inwardly, and it was one of the great joys of my life, one of the great lessons I learned. I learned that I could work hand in hand with nature, and make a useful contribution. Very few people have that experience, but to me, it was the real payoff for all my work.

Leichtman: It sounds as though you are talking about nature as a force or intelligence that is more than just the sum of all the plants and life on earth.

Burbank: Ultimately, it is an aspect of God—not an attribute, but an important aspect of God's life. Nature is the nurturing life force of creation, and without nature's nurturing love and intelligence, nothing could exist. The two of you might call it the Mother aspect of God. I called it nature, but it is really the creative intelligence of God. And humans are supposed to learn to work hand in hand with it.

Leichtman: Very good. Well, that suggests a question I've been waiting for the right place to ask. Were you aware of elementals helping you as you did your work? Did you see them?

Burbank: I didn't see them physically, but yes, I knew that there were forces helping me. I thought of them as the natural forces of the plant kingdom, rather than elementals, but that's what they were.

Leichtman: Well, the relationship between the elementals and the natural forces of the plant kingdom is so intimate that I suppose you could almost consider the one an extension of the other. It's not quite true, but it might well seem that way.

Burbank: Yes. I was not aware of elementals in the sense of specific entities, but I knew I was working with a force that was independent of the plant kingdom but helping it evolve. I regarded it as a divine force, and I was able to communicate with it and work in harmony with it.

Leichtman: Very good.

Japikse: Perhaps we should explain the role elementals play in nature for the benefit of those readers who might not be versed in esoterica.

Burbank: The elementals are God's caretakers or gardeners. They care for the nonphysical needs of the plant kingdom, providing nonphysical nutrients and creating a positive emotional climate in which the plants can grow. If I might put it in this way, they provide the plant with the motivation to grow and fulfill its purpose. They protect the plant from harmful conditions and try to provide it with the ideal conditions it needs. They serve as a bridge between the physical plant and the inner dimensions of life.

In a garden, the human takes on many of the duties of the elemental. But in nature, there is only the elemental to perform them. They make sure the plant kingdom is cared for.

Leichtman: Can they protect plants from the pollution of air and water?

Burbank: No.

Leichtman: That is entirely a human responsibility, then?

Burbank: Yes. The elementals are nonphysical.

Japikse: What is the nature of the interaction between humans and elementals? Can humans interfere with the work of the elementals?

Burbank: Oh, absolutely. The ideal is for humans to learn to cooperate with the elementals. But this does not often happen. In fact, humans often cause many problems and drive the elementals away from their gardens and cultivated areas. It must be remembered that elementals work on the emotional plane. If there is a human concentration of unpleasant or cruel emotions, it will drive the elementals away.

I have this picture in my mind of a very nasty person walking through a forest pinching elementals. *[Laughter.]* But that's not the way it happens. The bitter or unpleasant moods we sometimes carry with us are too coarse and negative for the elementals, and they withdraw from our company.

The elementals are responsible for maintaining the emotional environment in which plants live. This responsibility does not really extend to the physical. Humans are responsible for the physical environment of the planet.

Leichtman: Could there somehow be greater cooperation between the human and plant kingdoms in transmuting or transforming some of the environmental pollution that has gotten into the soil?

Burbank: There are natural forces that are able to deal with the problem of pollution—if they were permitted to deal with it.

Leichtman: Would that involve the plant kingdom?

Burbank: Of course. The mechanism for recycling the pollution of the planet is all there. And it does involve the plant kingdom and the elementals. The elementals cannot act on the physical plane, but they are in a unique position to aid the plant kingdom in purifying the environment.

Leichtman: What can we do to accelerate this cleansing? We know we have to cut down on polluting the planet, but what can we do to clean up the mess that already exists?

Burbank: I don't have specific suggestions for specific problems, but I will say this. The recuperative powers of the planet are enormous. If the pollution which has occurred is stopped, the damage that has been done will be recycled and the planet will recover. At this point, the damage is relatively insignificant, when compared to the recuperative powers of earth. I am not saying that we shouldn't be concerned about pollution or not take action to eliminate it—not at all. What I am saying is that we often fail to take into account the built-in planetary system for recycling pollution. What humanity really needs is a greater understanding of how these aspects of nature work, so it can cooperate better with them.

Having said that, let me comment briefly on what I consider the benefits of the movement to reduce pollution. I think it has done a great deal to foster a deeper sense of mutual responsibility for life on earth, and this is good. We all share the same soil and air and water, and we need to be more sensitive about how we have been befouling them. We need to learn to care more about the well-being of our neighbors—not just our human neighbors but also our neighbors in the plant and animal kingdoms.

Humanity has had a tendency to behave in highly individualistic, selfish ways, but this is not the way it should be. Eventually, we must learn to live in a refined state of cooperation. But we will never reach that point until we develop a strong sense of responsibility. The ecology movement has done a great deal to foster a deeper awareness of this need, and has promoted a deeper love for the earth and nature. All that is good.

Unfortunately, it has also tended to foster class hatred and a great deal of pettiness. It has attracted many good people—but also people who love to protest for the sake of protesting, not because they love humanity or the earth.

If humanity learns to become more responsible, then the ecology movement will have made a positive contribution. But part of this is learning that the earth has the capacity to clean up the mess we've made—if we will just let it.

Leichtman: Okay.

Japikse: Can you give specific examples of how human beings drive away elementals? You mentioned ugly moods. Are there other ways?

Burbank: That is the primary problem.

Leichtman: How about smoking cigarettes?

Burbank: As long as it was not indoors, the smoke would just blow away and that would be the end of the problem. Indoors, I can't imagine a smoke-filled room attracting very many humans, let alone elementals. *[Laughter.]*

Leichtman: How about someone who is very anxious?

Burbank: Of course. I would include that in my list of unpleasant moods. The principle is this: the basic link of communication with plants is on the wavelength of love, warmth, and cheerfulness. *Any* mood that drives these qualities away will also drive away the elementals and the plant will suffer. And quite frequently the reason plants suffer in adverse emotional climates is because the elementals have been driven away.

Leichtman: Yes. I have read stories about people who say that instead of talking to your plants you ought to yell at them and tell them, "I'm going to pull you up by your roots and throw you away unless you start blooming." They claim they get positive results.

Burbank: They do?

Leichtman: That's what they say. Are they just stimulating the survival instincts of the plant when they do that, or what?

Burbank: That would be one possible explanation, if they actually are getting positive results.

Leichtman: It doesn't sound like a compassionate thing to do.

Japikse: The plants are growing simply so they will get big enough to strangle the person. *[Laughter.]*

Burbank: Actually, it has nothing to do with what the human is doing. It is just a testimony to the strength of the plant to overcome obstacles.

Leichtman: It toughens them up.

Burbank: Yes.

Japikse: What about noise—specifically rock and roll music? Is that conducive to a plant's growth?

Burbank: Of course not. That should be obvious.

Japikse: It's obvious to us, but obviously not obvious to our next door neighbors. *[Laughter.]*

What is the impact of raucous music on plants?

Burbank: It won't kill them.

Leichtman: It actually has in some experiments, and in others, the plants will grow away from the stereo speakers in an effort to get away from the music.

Burbank: Let me point out that those experiments do not duplicate the conditions of the ordinary home, where most house plants would be exposed to rock and roll music. The love and care the plants would receive from the people living in the home would more than make up for the deleterious effect of the rock and roll music. But if the only element in the plant's environment was rock and roll music, then yes, I would agree that the plant would certainly suffer, and possibly die. But no plant is going to wither and die just from being exposed to rock and roll music. *[Laughter.]* There's a certain issue of balance involved here.

Leichtman: All right. Let me ask you this. How would you define the intuitive rapport you've said you had with plants and elementals? Would you call it clairvoyance?

Burbank: Not really. I didn't see the subtle bodies of plants or spirits. It was more a capacity to be in sympathy or harmony with the inner life of plants. I was able to merge my consciousness with theirs to some degree.

Leichtman: Did you read about any of these subjects or investigate practices such as mediumship or spiritualism?

Burbank: I did study the forms of meditation and the spiritual practices of my day, and settled on a simple form of meditation that seemed to help me contact what I viewed as "truth." I wasn't greatly enamored by voices and visions, though. While I thought there was probably some truth to it all, I didn't feel I had any business getting into that line of investigation. My intuitive impressions came at a different level of experience.

Leichtman: It sounds to me like the intuition usually found in geniuses. It's devoid of the flamboyant voices and visions and hot flashes and colors that so many people lust for—but even though it lacks all the glamour, it still works marvelously well.

Japikse: Yes. In reading one of the biographies of your life, I was intrigued to find that your family had at least a passing interest in spiritualism and talking to spooks as you were growing up. Did that in any way help you become more aware of or attuned to the subtle dimensions of life?

Burbank: I don't think those family interests did anything to make me more attuned, but I will say this—my family environment did not in any way drive my awareness of the subtle dimensions out of me, and that alone is noteworthy. I will admit that there were quite a few kooks in my family, but I was more wrapped up in other things. I wasn't negative about these interests, but they weren't that much of a positive factor in my development.

Japikse: There seemed to be times in your life when you did the equivalent of spiritual healing. If I remember the story correctly, some children in Santa Rosa had some difficulties physically and you were able to heal them by laying on of hands.

Burbank: It was never my intent to be a healer. It was just a simple extension of the work I was doing with plants. I figured that if I could communicate with the inner essence of one form of life it should be relatively easy to do the same with all forms of life. It was a natural thing for me to do, and yes, I did have some limited success. The few times I did this have been blown out of proportion. But it is true that the same basic principles of healing can be used with all forms of life.

Leichtman: You also seemed to have a capacity to absolutely dazzle people and turn them into loyal and adoring fans—apparently without intending to. Today we would call it charisma. Of course, it seemed to work against you as much as for you, and those who opposed you were as steadfast as those who adored you. My question is this: what gave you this special quality?

Burbank: I don't know. It probably has a lot to do with the inability of most people to understand and perceive what I was doing. Most people thought what I was doing was magical—that somehow I had magical powers that were not available to all. And those who loved magical powers adored me. But others simply became jealous of the powers they thought I possessed. So I guess the source of my charisma or mystique was the impression most people had that I was doing something extraordinary. Of course, to me it didn't seem extraordinary at all, because I had spent my whole life doing it.

The different reactions of people were quite amusing at times. The reactions of jealousy got to be annoying after awhile, of course, because it was a bother to have to put up with that all of the time. But I understood that these people had no comprehension of what I was doing; they were not able to look at life from the same perspective I had. Conversely, the reactions of adoration could also become excessive, but I knew they were looking at my work and not really seeing what it meant. They were looking only at the physical results I was producing and not understanding that all I was doing was working in harmony with natural law. I was flattered and grateful for their adoration, but most of it was misplaced.

Leichtman: Of course, a lot of the criticism directed at you came from the scientific community. There were various experts who felt that what you did was not all that unusual and that your claims were exaggerated, because it wasn't scientific.

Burbank: A lot of that criticism came because I always tried to do things simply. I was interested in results, not scientific papers, and so I didn't document my methods as thoroughly as some scientists might have liked. They didn't have the imagination to fill in the dots and recreate my experiments on their own, so I was accused of running experiments that other scientists couldn't duplicate.

Of course, I often said that I was not doing anything that other people couldn't do. But to duplicate my work, other scientists would have to apply themselves and study life itself and learn from their experiments, as I did. And that's what they couldn't—or wouldn't—duplicate. They were only interested in duplicating my results, not my methods. And that's why my work seemed magical or unscientific. My approach to plant life was different than the approach usually taken. I was aware that working with plants was more than a mechanical process. I regarded plant life as *life,* and not just the object of scientific inquiry.

I want to be as clear about this as I can. Much of my work with plants *was* mechanical—knowing how to graft and accelerate the growth of plants and select good strains. These were the mechanical details of my work, and they can be mastered by anyone who is observant and diligent. But to be truly creative, you have to have more than these mechanical skills. You have to approach plant life as a form of life, and understand it as such.

This is where the materialistic focus of scientific experimentation today is such a great hindrance to human advancement. The person who looks only at the finite, measurable, tangible, and quantifiable elements of a phenomenon is going to miss the secrets he wants so desperately to uncover. This may sound paradoxical to the scientists who will read the book, but it is true. The explanation of *life* cannot be found in tangible measurements.

In my day, I was thought of as very unscientific. The university professors that came out to look at my methods wondered how I got anything done at all, because I was sloppy, careless, and unconcerned with scientific purity. For example, I didn't isolate all of the variables in my experiments. But I wasn't interested in that. I was just interested in getting results.

I'm sure the criticism would be even worse today.

Japikse: Well, you didn't have a university grant, so you must not have been doing scientific work! *[Laughter.]*

Burbank: That's right. Much too much is made of degrees and the fact that you are associated with a university or a laboratory. To me, that's all absurd. I wouldn't have been able to accomplish a fraction of what I did if I had been supported by a university grant or conducted my experiments in a laboratory setting. To me, all of nature is the laboratory in which we must conduct our experiments.

Leichtman: I guess we don't have to ask you your opinion of "scientific horticulture" then. *[Laughter.]*

Burbank: Well, as I said before, too many experts believe their expertise gives them the prerogative to set down rules. But in dealing with plants, it is nature that sets the rules. You can't afford to take premises as facts unless you thoroughly appreciate the way the life force is acting through the plant. Let me draw an example from spiritual healing. You might have a very good healer who is able to tap the spiritual life of the people who come to him and use it quite successfully to heal their ills. This is the real healing work. But sometimes healers fall into certain patterns. Let's say this fellow has developed a style where he squeezes the right shoulder three times and kicks the person in the back of the knee. *[Laughter.]* There would be any number of people who would think that the secret to healing is to squeeze the right shoulder three times and kick people in the back of their knees. *[More laughter.]* There would even be some who would try to measure the pressure exerted on the shoulder and the timing of the kick! *[Guffawing.]* But that has absolutely nothing to do with healing.

And that's my opinion of most "scientific horticulture."

Leichtman: People like that are just pseudoscientists.

Burbank: But they pass for the real thing. I know, because I had to deal with these types. And it was always amusing to see how much emphasis they placed on their hard and fast rules, because to me it was a very natural process. But they would come to study my methods and watch me perform a certain procedure, and then they would ask, "Why did you decide to choose those two plants to graft," expecting me to give them a hard and fast rule that could be put in a textbook so that non-horticulturalists could learn it, too. And my only answer would be that they were the right two plants.

I guess they wanted the equivalent of painting by numbers.

Leichtman: I suppose genuine scientific work has much in common with creativity. I think you just described the process of creativity.

Japikse: Yes, especially the discovery aspect of it. Did you think of yourself as a discoverer or a scientist or an artist?

Burbank: I thought of myself as a person who loved plants.

Leichtman: What would you consider to be the most important characteristics of genuine scientific inquiry?

Burbank: We've already discussed some of them, but let me add this. One of the most important characteristics that I had was a nonstop, insatiable curiosity. Even as a child, I was enthralled by nature and wanted to learn as much about it as I could. There is just so much to observe and so much to learn!

The scientist really needs to regard nature not so much as a laboratory but as a textbook—the best textbook there is about life. And it is our curiosity which inspires us to read this magnificent textbook. I guess I was lucky to have this characteristic as a child, because I never stopped observing and trying to understand the phenomena of nature. And this is how the good scientist learns—he learns from the phenomena of nature. Laboratory experiments have their place, but the truly creative scientist is never limited by the laboratory.

It was this intense curiosity that taught me how to look for the subtle clues and signs that let me tap some of the secrets of nature. There really aren't any secrets, you know. Nature is there for everyone to see; the laws of nature are there for all of us to observe. It just takes someone with imagination and perseverance and perception to recognize them and understand their meaning.

I might add, by the way, that an uncle of mine was a professor who helped me to appreciate that there *were* answers to my inquisitive questions, and there were explanations to some of the things I had observed but not yet been able to figure out. This gave me a very good role model for my later work, and was also a great intellectual stimulant. It taught me that my curiosity could be satisfied and that other people had developed their minds to the point where they could answer the questions I had.

There was nothing magical about what I did; to my mind, it was highly scientific. The fact that I could see things other people couldn't see and hear things other people couldn't hear and know things other people couldn't know didn't mean I was working magic. It just meant that I had refined my skills of curiosity and careful observation. I was *looking* for the understanding that other people couldn't be bothered with.

I was intuitively guided, but I don't want anyone to think I got all of my ideas intuitively. The intuition helped me understand what I saw, but what I saw was the result of my intellectual curiosity. My capacity to see that a certain seedling or sprout had the potential I was looking for, and the rest of the batch did not and could be destroyed, was not a psychic talent; it was a skill of careful observation, nurtured in my childhood by my curiosity.

A second characteristic the scientist needs is the awareness that there is more to life than the physical forms he observes. He needs to understand that the real life of plants and animals and all of nature is an intangible, nonphysical quality. The true scientist—and this is going to upset a great many pseudoscientists—has to be willing to deal with aspects of life for which he has no direct tangible evidence. If he isn't, he'll never be able to understand the meaning of what he is able to prove. He will just end up being a materialist.

When I dealt with plants, I always worked with the knowledge that the form was just one part of the whole life of the plant. I knew that the development of the plant was guided by intelligence and that I could contact and be guided by this intelligence, too. I couldn't prove this, however, so I didn't talk about it much. I had enough trouble as it was without adding to it! But I couldn't have done what I did if I had not had this understanding.

Japikse: As I've been listening to you talk, it's occurred to me that this awareness that there is more to life than just the physical form is an idea that would be useful to the average person as well as the scientist.

Burbank: Absolutely. And the average person can't prove it any more than the scientist, but he can sense that it enriches his life. Let me give you some examples. We could use the example of a secretary who performs routine office work day in and day out. This secretary always executes her tasks competently and efficiently, but every now and then she hits a peak level of performance where she touches a level of inspiration and efficiency which transcends her normal competence. What produces these peak levels? The work she is doing is not different than her ordinary routine, but something inside her is lifting her skills and understanding to a higher level.

We all experience moments like this. Perhaps we become more joyful or optimistic or tolerant than we ordinarily are. And we should view these moments as lessons which teach us that there is more to life than the tangible, physical conditions we are familiar with. The creative person is one who learns that not only do these greater potentials exist within himself, but also within the whole of life around him. And he teaches himself to relate to and work with these greater potentials. He can't prove to anyone else that these greater potentials exist, but there have been times when they have peeked through in his life. And having once recognized them, he learns to call upon them again and again.

Now, it's this capacity to be aware of the inner potentials of life that really opens the door to creative discovery and activ, ity. Let me give you a silly example. Let's suppose that when Beethoven was first inspired with the melody for *The Ode to Joy,* it didn't occur to him to write a choral symphony. Instead, he decided it would make a good hum. So he wrote out the melody and hummed it for his friends. Then it occurred to him to score it for the piano, but before he could do so, he changed his mind and decided to write it for a string quartet. And in this way, he just kept on expanding his idea of how to treat this melody, until he decided to have it performed by a sym, phony orchestra, and then a symphony orchestra with four vocalists, and finally a full orchestra and chorus. The inner life of *The Ode to Joy* would be the same, of course, no matter how it was performed. But Beethoven had a choice whether to make it a simple hum or the final movement of the Ninth Symphony.

It was exactly this kind of realization that made so much of a difference in my work. Because I was aware that there was more to a plum or a potato than the outer form, I could recog, nize that each of the different varieties of this fruit or vegetable was a single expression of a rich and complex inner life. And so, if I worked on a large enough scale with individual plants, I would be able to find the characteristics or traits I needed to produce a new and better form. But I wasn't improving the inner life in any way. I was just finding new ways to give it expression on earth.

It was my work to collect desirable growth characteristics and qualities, looking at thousands and thousands of seedlings from a single species, and then produce new generations in which these qualities would be strengthened. But I was not just following my own course. I was working in partnership with the inner life of the plants, and I knew I had an obligation to let it guide me.

The true scientist learns to do this.

Japikse: How could the ordinary gardener relate to the inner life of plants?

Burbank: Well, the ordinary gardener should try to understand his responsibility to the plants. To some degree, he is taking on many of the functions of the elementals—he is becoming the caretaker of these plants. So first of all, I think he should learn what he needs to know about caring for these plants physically. It does no good to love the inner life of a tropical plant if you plant it outdoors in a wintry climate. So you have to learn some facts—whether a plant needs shade or sun or both, how deep the seed has to be planted in the soil, what kind of fertilizer to use, and so on. If you don't know these things, you can't care for your garden adequately.

Beyond that, the ordinary gardener should understand that plants are responsive to their environment, along the lines we've already discussed. They know if they are loved and appreciated or seen as a burden. Now, I don't want anyone to get the wrong idea. Plants don't have the peevish kind of emotional reactions that humans do. They are sensitive to the emotional environment, but not really in the sense of having feelings. They won't get mad at an angry human, for instance. But they do not thrive in that kind of environment. So, anyone who wants to do a good job of raising plants would feed them water and fertilizer and make sure they are planted in the right soil, but he would also feed them kindness, affection, and goodwill, to nurture their inner life. And as he's working with the seed or the seedling, he should think lovingly of the way this plant will look when it is in full flower, and the beauty it will add to the garden. Plants respond well to this kind of loving expectation, because it helps strengthen their inner life.

Japikse: What impact does the inner life of plants have on humans? Let's say your whole house was filled with plants?

Burbank: Most of the effect is indirect. It is real, but indirect. Someone like yourselves can intuitively tune in directly to the inner life and be aware of the plants leaning forward or bursting with joy, and that enriches your life. But for most people, the impact is indirect. The inner life of the plant radiates its warmth and beauty into the home environment, and this helps provide a healthy emotional atmosphere.

House plants are usually very beautiful physically, and often have a fragrant perfume which is uplifting as well. But there is a counterpart to that beauty and fragrance at the inner levels which is really the substance of the high emotional qualities. At the inner levels, the flowers will often radiate subtle qualities of joy, peace, affection, optimism, or love. And this can have an uplifting impact on many people.

Leichtman: How does the plant build up this strong, radiant force of joy or optimism?

Burbank: It is inherent in the nature of the plant to do this. It is part of the evolutionary work of the plant kingdom to accumulate positive emotional qualities, as I said earlier. And so individual plants are designed to do this, too. You might as well ask, "What makes the sun shine?" Well, it is designed to shine. And plants are designed to accumulate and radiate positive emotional qualities.

Plants just love to grow.

Leichtman: Yes.

Burbank: They love to be themselves. If you plant a green bean, it's greatest joy will lie in becoming a mature green bean. It doesn't have any choice in the matter, of course; the innate design of the green bean has already decided the direction in which it will evolve. But as the life force of the bean moves towards its destined perfection, it does generate a measure of joy—the joy of fulfilling purpose and design. And this joy becomes a part of the inner life of the plant.

Leichtman: We've already talked a good deal about the

evolution of the plant kingdom, but is there anything else you would like to add about the process of growth and nurturing growth, either in plants or in humans? Or what you learned about the capacity of humans to nurture growth? This seems to be an area of great interest for you.

Burbank: Yes, it is. To me, it is obvious that all life forms in nature carry with them an innate design for their evolution. Whether it is a buttercup or a butterfly or a human being, there is an inner intelligence and life force which oversees its evolution and growth. This life force is not the only factor influencing growth, however. Nature also plays an important role, providing circumstances and conditions which promote growth—weeding out the bad seed, strengthening the good characteristics in a life form, and so on.

Man can also participate in this process. I found that I could often enormously accelerate the natural processes of growth by intervening and playing an intelligent part in weeding out, enriching, and strengthening the seed. I always worked within the bounds of the inner design, but I sometimes was able to accomplish in a few years what might have normally taken thousands of generations to achieve naturally.

After a while, it became clear to me that the inner intelligence was just waiting for a better form to be created so it could express more of its life and beauty through it, whether it was a rose or a daisy or a plum tree. And I occasionally had the insight that I could have gone on improving any one variety almost indefinitely, because the inner life would not have been exhausted. But of course at some point I had to stop and realize that this was good enough and make it available to the public at that point.

The lesson to be learned from this is that human beings really can't put any limit on the potential of growth, whether it is in our own children, in the plants in our garden, or in any species of animal on the planet. And once you know this basic truth, you become devoted to doing what you can to accelerate growth on earth. Because you are guided by the basic impulse to grow. You make mistakes, of course, but this impulse keeps you headed in the right direction.

Leichtman: It sounds almost as though you are describing the fourth-dimensional force of growth.

Burbank: I am. To understand growth, you have to see the larger context in which it occurs. You have to understand the inner forces motivating it. Now, this is not as complex as it might seem as we sit here talking about it. Any parent can easily tap into the fourth-dimensional force of growth for his children simply by learning to recognize that there is a mature, wise adult lurking somewhere in the tiny, immature child. This wise, compassionate adult has not become manifest on the physical plane yet, but it is there waiting to appear. And it's not just a hope or a dream; it is a real presence in the life of the child. And the parent can be guided by this wise adult within the child as he or she helps the child mature.

This is the way I worked with the fourth-dimensional force of growth in plants, too. I recognized that somewhere within this plum tree with sour fruit and big pits there was the essence of plumness. *[Laughter.]* My challenge was to be guided by this essence of plumness and restrain the characteristics which produced the sourness and the big pits and strengthen the characteristics which would make the plum a better fruit. I was doing the counterpart of the parent, who trains his children to be self-restrained, affectionate, intelligent, and mature.

This is what it means to work with the fourth-dimensional force of the inner life. The inner life is greater than the form through which it appears. It contains the pattern, the power, the substance, and the real character of the form.

Japikse: Just to nail this down, then—the wise parent invokes the mature seed within the child, rather than imposing his or her concept of what the child ought to become?

Burbank: Well, yes and no. That's true up to a point, but obviously good parents do impose their values and good habits on their children to some degree. Sometimes they do this directly, by saying, "Do this and not that," and sometimes they do it indirectly, by being a good role model. And this is appropriate. But you are right, the parent should always be guided and led by the inner life of the child, and work to draw out the child's inner wisdom and goodness.

As I've already indicated, in my work I did things for the plant which the plant couldn't do for itself, and this is the ultimate expression of stewardship—to help a life form, be it a plant or a child, to go beyond its scope of evolution. And it is part of the spiritual role of humanity on this planet to learn to nurture the growth of all forms of life on earth.

Leichtman: So the true role of science is not just to discover what it can about life, but even to go beyond that and find out what we can do to enrich life and facilitate its growth.

Burbank: The true role of science is to understand life and comprehend creation so we can participate in it and add to it. A scientist is meant to be a steward of life forms. He is charged not just with measuring and quantifying data and writing papers, but to get busy fostering the work of creation.

Leichtman: I'm eager to ask a question on a slightly different line. During your lifetime, many giant intellects came to visit you. I remember seeing one photograph where both Thomas Edison and Henry Ford were visiting you at the same time. What was the tie there? Were you able to be in tune with their genius in the same way you were in tune with the intelligence of plant life?

Burbank: Sure. We had quite a group and were quite close. And the bond that tied us was an understanding of each other and what we were doing. There was a sense of camaraderie which I will always remember because we were not in the least bit jealous of each other. We all shared a similar love for life, and respected the work each one of us was doing. Even though we were working in different areas, we all derived great joy from the work we did, and it was this capacity to enjoy our life and work that created such a strong bond.

I might point out that we were all human. We all had personality flaws and human frailties—our own individual eccentricities. But what we shared was our common bond of joy.

Leichtman: This seems to be common among people who play very important roles in society—they have the capacity to respect the genuine contributions of others. Let me ask you about these personality frailties. Do you think they are inherent in the creative or intuitive genius, or do they simply get greater exposure than the same problems in ordinary people, because you are in the public eye so much?

Burbank: I always held the theory that if a person is involved in creative work and fulfilling his destiny, everything else is relatively unimportant. It doesn't really matter if I have eccentricities or not—if I wear one black shoe and one white shoe, or anything else—if I am in tune with my own destiny and am fulfilling it. Sometimes creative people work with such intensity in their field of endeavor that they approach the rest of life with more lightheartedness than other people do. It is their way of relaxing from the intensity of their work. And so they may end up seeming childish or frivolous at times. But they really aren't.

Leichtman: In other words, your love of plants wasn't really a sublimation of an Oedipus complex? *[Laughter.]*

Burbank: Oh, I am sure it was. And Henry Ford was a real crank. *[More laughter.]*

Leichtman: Well, since we've stumbled into this line of questioning, what is it that makes a Luther Burbank? We've talked some about your childhood, but certainly you are much more than just the product of your New England upbringing. What forces, experiences, and inner guidance turned out the genius in Luther Burbank? Did you have a past life as a parsnip or something? *[Laughter.]*

Burbank: I think it is fair to say my life as Luther Burbank was the culmination of many lifetimes of study and growth—as well as study and growth on the inner planes between physical lifetimes. A life such as that doesn't just happen by chance. I had spent many, many lifetimes before my life as Luther Burbank working with the plant kingdom and developing my skills and understanding, so that I would be able to play the part I did.

Leichtman: I take it you are part of a large group working together in this way, but on a vast scale.

Burbank: Sure. There is a great division of labor in the sky. *[Laughter.]*

Leichtman: Well, what are you doing now?

Burbank: I'm very busy. After a whole life of being myself, I needed a short vacation. But then I immediately continued my work on the inner planes.

Leichtman: What does that consist of?

Burbank: I work from time to time with people in the physical who are working experimentally with plants. I try to help them become more aware of the inner life and potential of plants.

There is a lot of new work being done in learning to use plants for both psychological and physical healing, and I am involved in that too. Some of this has already become known to the public, but most of it is in a long-range developmental stage and won't come to fruition for several decades. But there are some interesting experiments going on.

Japikse: Are you working as intensely as you did in the physical?

Burbank: Much more so.

Japikse: Still doing thousands of experiments at the same time?

Burbank: Oh, tens of thousands. Well, it's not quite like that. Life here is much different than on the physical plane. You aren't limited by a dense physical body. Experimentation is a whole different process, because in many cases you don't have to actually conduct the experiment. You just look at the variables and you know how it will turn out. That's the advantage of working fourth dimensionally, after all. You can see what is optimal and the conditions which will produce those results, and you can create those conditions almost instantaneously.

The experimentation that we do engage in is more along the lines of experimenting with long-term evolutionary outcomes. We start with a basic proposition or plan and see what needs to be done first and what needs to be done next in order to bring it to fruition.

Japikse: Well, let me ask this. In the physical plane, we can tangibly see the growth in a plum tree or potato. What is the appearance of growth at your level? How do you know something is actually growing?

Burbank: Do you mean the subtle forms of plants?

Japikse: It might involve that, but wouldn't have to be limited to it. How does growth occur at the inner dimensions?

Burbank: Well, it's not easy to describe that without sounding vague, but I'll try. Let's use the example of the growth of an idea. It grows by gathering to itself the life force of its own level, until it builds up to such an intensity that it begins to expand outward and into the lower dimensions of life, until eventually it achieves expression in the physical plane. And this can be observed as the idea becomes more powerful, more structured, and more responsive to purpose. It also gradually takes on more concrete detail. All of this would be the appearance of growth.

Japikse: So you are saying that growth is an inherent aspect of life?

Burbank: Yes. Growth comes from within. There is an inherent design and life force which stimulates and guides the growth we see in form.

Japikse: Is there any way the people reading this interview would be able to attune themselves to the work of the group on the inner planes you were mentioning a few minutes ago? I'm talking about people who really love plants and would like to be more responsive to the inner work being done.

Burbank: The best recommendation I can give is that they should spend a great deal of time becoming thoroughly acquainted with one type of plant or species of plants, until they know all about it—how it thrives and grows and regenerates and so on. I am not recommending this as therapy to escape personal problems, mind you, but for those people who truly love plants and want to work with the inner design for plants. As they become thoroughly versed in the physical characteristics of this type of plant, then, they should strive to understand the subtle dimensions of it and attune themselves to the consciousness of the plant. For starters, this can be done just by realizing that this plant is alive at inner dimensions and is part of the evolving life of the planet. And they should try to sense a kinship with the plant, and know that it has a divine essence just as they do. Practicing an effective form of meditation would help build this rapport with the consciousness of the plant. And as you start to work at that level, you will automatically come into contact with the inner forces working to assist it in its development.

Leichtman: There are some exercises in Evelyn Underhill's book *Practical Mysticism* that might be useful in this regard. They involve trying to imagine what it would be like to be a rose or a tree.

Burbank: For the person who doesn't have any interest in meditation, sincere love and desire to help plants grow is all that is really necessary—that and the attempt to expand his thoughts and appreciation of the plant beyond the physical level.

Leichtman: There are some people who play soft, gentle music to their plants every afternoon. I remember seeing one fellow who spent an hour every day walking up and down his garden rows playing the flute to his plants.

Burbank: I'm sure they just loved it.

Leichtman: Yes.

Japikse: Are you familiar with the Findhorn project?

Burbank: Yes.

Japikse: Is this a productive way of working? Is this one of the projects under the auspices of that group in the sky?

Burbank: It certainly is a viable demonstration of what can be done and how communicating with plants can be beneficial. If you are looking for an indication from me whether or not they are achieving what they say they are, well, yes they are.

Leichtman: Its character has changed somewhat since the beginning, hasn't it?

Burbank: Well, most organizations go through that process.

Leichtman: I understand there have been some rather radical changes. In any event, I don't believe they have made much effort to develop new varieties or hybrids. Their success has been largely in the field of raising extremely large vegetables and flowers.

Burbank: The location has a lot to do with that.

Leichtman: So I understand.

Burbank: The results they have achieved are probably more due to the area than to the consciousness of the people, although it is a combination of both.

Leichtman: Well, what is so unusual about the area? I understand the soil is very sandy and it is very cold. It's in northern Scotland, after all.

Burbank: It isn't the physical area that makes the difference.

Leichtman: That's why I'm asking.

Burbank: There is a strong vortex of high-quality emotional energy at that spot which makes it very conducive to raising oversized plants.

Japikse: Was the vortex there before Findhorn was started, or did they create it?

Burbank: It was there before they arrived.

Leichtman: Was it left over from Atlantis?

Burbank: No.

Leichtman: It just happens to be there?

Burbank: No, it was put there.

Leichtman: Who put it there?

Burbank: Now, I'm not going to tell you that. *[Laughter.]*

Leichtman: Awww. Well, how can we get one?

[More laughter.]

Japikse: It would make a lovely Christmas gift.

Leichtman: Yes, I want one, too. Was it created by the deva kingdom?

Burbank: Yes.

Leichtman: So it is not really subject to our human whims or desires.

Burbank: No.

Japikse: Was there a similar vortex in the part of California where you worked?

Burbank: No. It wasn't necessary for my work. This is why I say the Findhorn work is more an experiment involving the area than the people participating in it.

Leichtman: If a group of people who were all devoted to nurturing life gathered together and focused their group intention along the principles you have expounded in this interview—reverence for life—could that in time invoke these angelic forces?

Burbank: Of course. The forces of the deva kingdom as well as elementals are very quick to respond to any positive invocation from the human kingdom.

Japikse: What do you think of the work John Ott has done in his experiments with different qualities of light and their effect on the plant kingdom? I am referring to the research he's reported in the book *Health and Light*.

Burbank: John Ott has done brilliant work and, I must say, genuinely scientific work. The pseudoscientists of today have contempt for him, as they did for me, because he didn't exclude all of the possible variables involved in his experiments. But he has produced results, and that's what counts. Here is a man who simply began observing what most everyone else missed—that there is a big difference in quality between artificial light and natural sunlight. And he has proven that we need to duplicate natural lighting conditions as much as possible, whether it is in our homes or offices or in our greenhouses.

I certainly tried to do that for my plants—I tried to create optimal growing conditions in every way possible. And as a result, my flower beds and garden plots were amazingly productive. It wasn't amazing to me but it was amazing to others.

People need to appreciate the work John Ott has done

more than they do, and appreciate his style of working, too. And they ought to read *Health and Light*. It will get them thinking. You know, the quality of light and radiation we are exposed to is only one of many aspects of life which affects our growth. And if the quality of light can drastically affect our growth, as Ott has shown, then maybe we ought to be wondering if other variables can affect it, too.

Leichtman: What other factors?

Burbank: Well, certainly sound would be one, and our emotional environment would be another. We've already discussed these things to some degree. In planting a garden, even the kind of plants you put next to each other can affect the growth of the garden. And the kind of fertilizer you use can also have an effect. Should it be artificial, or the product of an oil well, or natural? These are all factors to explore.

Leichtman: Is there anything else you wanted to ask?

Japikse: No.

Burbank: Well, let me make one final statement. I never could have made the contributions that I did unless I had gone an independent course. The creative person must learn what he can from schools and the traditional approaches of society, but then he must go out and create his own niche and style of working. And he must pursue his own work in spite of the criticism and protests and contempt and even indifference that are bound to arise. Because the further your work develops, the more the criticism increases. I was fortunate to receive as much adulation and praise as criticism, but I did attract an awful lot of criticism. Anyone who blazes a path that goes beyond his predecessors will, because people are not going to understand you or your work. You have to be strong enough to carry your work on your own shoulders; you can't be dependent on the approval of others.

And this is where it is absolutely critical for the creative person to understand the nature and principles of growth. You must know that growth comes from within. You can't depend on a chorus of friends and angels to sing praises to your creative efforts every morning to get you going. *[Laughter.]* You have to be able to get up and get going without any external encouragement, and to do that, you must find the inner wellspring of life that can sustain and encourage you.

Once you discover that inner wellspring, you will find that each day provides you with fresh evidence that your work has value. And this evidence sustains you even when the world does not approve of what you are doing, because you know the inner life is guiding you. If I hadn't had this realization as a constant experience, I never would have developed my first potato.

Leichtman. That certainly puts a fresh perspective on a theme we have often discussed in these interviews—the need of society to change and become more supportive of its geniuses and creative people. And it suggests that if nothing else, as the creative person learns to deal with criticism he also learns to become more reliant upon the inner life.

Burbank: Yes. Now, keep in mind that I am not talking about people who just become sensitive to criticism and withdraw from their creative endeavors and become neurotic or eccentric. Some of these people are just protecting themselves from injury; they aren't doing anything creative. But yes, truly creative people have to recognize that they must be able to deal with criticism—and the best way to do this is to become inwardly directed.

Japikse: It can certainly be an impetus that increases the reality and immediacy of the inner life.

Burbank: Yes. I will make one more comment about the inner life and then return to it. *[Laughter.]* The more the creative person becomes attuned to the inner life, the easier his work becomes, because that is the seat of creativity. There are all kinds of great ideas and marvelous new developments in the inner life that have never appeared on earth but are waiting to be discovered and written down and experimented with. These new ideas don't have to be invented, any more than I had to invent the potato. The potato was already there. My work was to develop a better potato. I helped it grow. And that is the true creative work of humanity—not to invent totally new expressions of life, but to play a creative part in helping the life that is already here on earth grow and become better.

No one invents the inner life. It already exists. It is real. And it guides and stimulates our creative work. The person who can learn this can become a true creative worker.

I think I will close on that note.

Leichtman: Thank you for a very good interview.

Japikse: Yes, thank you.

Burbank: It was my pleasure.

SIR OLIVER LODGE RETURNS

The great value of the scientific process is that it leads us to new discoveries about the world we live in—discoveries about nature, physical laws, and our own selves. Whenever science helps us make the discoveries that lead us to a greater understanding of life in its fullest scope, it serves a noble purpose. When it narrows its vision, however, and becomes too preoccupied with only tiny fragments of life, it fails to honor its true mission. Many scientists, unfortunately, believe that science must limit its investigations to the physical aspects of life alone. The issues of human psychology, life after death, psychic phenomena, and the impact of nonphysical forces upon our life, they profess, are somehow outside the realm of *their* science. They scorn scientists who have the courage to investigate these important questions, and a priori reject their data and conclusions, on the grounds that such research jeopardizes the reputation of science. Of course, it is their attitude which most injures the true purpose of science, as it intimidates all but the bravest scientists from exploring the vast, rich dimensions of the invisible portions of life.

One scientist who understood the dangers of the materialistic approach to science—the idea that science must concern itself only with the physical aspects of life—and worked to counter it was the British physicist, Sir Oliver Lodge. One of the principal contributors to the development of the radio, Sir Oliver also was a major precursor of Einstein's theory of relativity. He held the chair in physics at University College in Liverpool from 1881 until 1900, when he became the first "principal" (president) of the new Birmingham University—a post he served until 1920. He was knighted in 1902 for his contribution to physics. In 1913, he served as president of The British Association for the Advancement of Science.

Throughout his career, Sir Oliver was also an active investigator of the nature of psychic phenomena, life after death, and the spiritual implications of science. Far from considering these aspects of the "hidden side of things" incompatible with science, he saw them as integral parts of any genuine scientific study. He knew from his research into radio waves and other electromagnetic radiations that science is really a study of energies and forces, many of which are beyond our current understandings. So, any activity that broadens the involvement of science in life must be helpful.

Sir Oliver first became acquainted with the investigation of psychic phenomena through his membership in the Society for Psychical Research, which he joined in the 1880's and of which he was president twice. This society was composed of many of the leading figures of England, and to a large extent founded the modern science of parapsychology—the investigation of psychic ability, life after death, and other phenomena associated with the invisible dimensions of life. His interest became more personal when one of his sons, Raymond, was killed during World War I. The contact he succeeded in establishing with the spirit of Raymond, through various mediums, became a classic in the history of mediumship. Once the original contact had been made and the Lodge family was reassured of Raymond's existence, Sir Oliver and Raymond continued with their mediumistic contact, both to reliably prove that such communication can indeed be genuine, and to gather information about the conditions of life after death. Raymond became the principal investigator in the latter work, from his new perspective as a spirit.

Both in his conventional scientific work and his investigations of the finer realms of life, Sir Oliver demonstrated an important element that is often missing in scientists. His efforts were primarily guided by a sense of humanistic concern, rather than by the opinions of his peers or mere scientific tradition. He considered his scientific work as a way to make a significant contribution to the knowledge of mankind and always proceeded on that assumption.

This is a relevant example for all scientists of today to follow, whether or not they are also interested in psychic phenomena. But it is especially of value to those scientists who are helping to establish the fledgling science of "parapsychology." Too many of these people have narrowed the scope of their research to the point where it is almost without value. They confine themselves to measuring laboratory phenomena and compiling statistics, rather than *viewing* the effects of mediumship, telepathy, clairvoyance, materializations, and life after death.

One of Sir Oliver's greatest contributions as a scientist is that he recognized that the element of *human consciousness* is of far more importance to the study of psychic phenomena than physical measurements. And yet, his example has been all but forgotten by today's new breed of psychic investigators, who drool over statistics and mechanical measurements of psychic abilities. Sadly, this is a great contradiction; trance mediumship was reintroduced in the 19th century for the purpose of offsetting the gross and militant materialism that afflicted science and most of society at that time. The study of mediums and psychics, therefore, was meant to lift our eyes away from the purely physical and help us become more aware of the spiritual and mystical elements of life. Now, however, the very field which was designed to neutralize the forces of materialism has been almost completely overwhelmed by materialistically-oriented scientists posing as parapsychologists. In effect, there has risen a new class of "esoteric" materialists, who cheapen and trivialize the value of psychic phenomena by investigating only those psychics and mediums who are themselves trivial—people who can bend a key, snoop into a competitor's laboratory, find lost objects, affect electrical instruments, and so on. These physical phenomena are sometimes useful, but intelligent people should remember that they are indeed *physical* phenomena. Psychic phenomena, by contrast, can only be understood by the investigation of its nonphysical aspects.

Psychic abilities are designed to make us more aware and knowledgeable of the aspects of creation that complement the physical plane. Of course, these abilities and the information gained by their use can greatly enrich our knowledge of the physical plane, but we should never forget that we are meant to aspire to a greater and more inclusive knowledge of all of life. We are not meant to cheapen knowledge!

It was with this belief that Sir Oliver labored. He, too, was interested in validating the legitimacy of specific psychics and mediums—indeed, his experiments and recorded experiences incontrovertibly demonstrate the scientific fact that the human personality survives death and can be intelligently contacted. Those modern "scientists" who continue to doubt this fact would be well advised to research the rich legacy left by Sir Oliver. But Sir Oliver also went beyond mere proof and studied the implications of survival—what it reveals about human consciousness, the process of inspiration, and the relevance of survival after death to physical life. It is sad that these landmark contributions are usually given only lip service by the investigators of psychic phenomena today.

Indeed, many modern parapsychologists actually scorn and reject the possibility of studying life after death, as though it is beyond the scope of science. In doing so, they miss an extremely important point: in order to meaningfully study the paranormal field, it is necessary to appreciate the fact that there are independent realms of creation apart from the physical plane. The human spirit is not dependent upon either the physical body or brain for its existence; in fact, consciousness can exist totally apart from the physical body and nonetheless continue to influence the physical plane.

This is a fundamental thesis in the process of bringing heaven to earth—of increasing our awareness of the fullest dimensions of life and acting upon that awareness for the benefit of mankind. It is for this reason that it is disappointing to see so many researchers believing that they can discover the real elements of consciousness by measuring brain waves, studying brain chemistry, evaluating enzymes and hormones, and attempting to change abnormal behavior with complex nutritional and chemical strategies. Consciousness does not exist in the brain or in our hormones! It is entirely nonphysical. Thus, if we are to really understand consciousness, we must dust off Sir Oliver's methodology and findings, and once more apply the spirit of his work to our continued work—whether we are a student of human nature, a psychologist, a psychiatrist, or just a dedicated parent who wonders why our children behave the way they do.

In the interview that follows, Sir Oliver comments about current scientific investigations into psychic phenomena and life after death, and makes suggestions for getting the whole effort back on a more useful track. He encourages investigators, for example, to search out talented and competent psychics who are quietly helping people cope with their lives, rather than the flamboyant psychics who may produce more spectacular effects but who are really aberrations more than they are talented. He also makes a number of forceful comments about common psychic and mediumistic practices—especially the practices of the Spiritualist Church, which his work helped establish.

At one point in the conversation, Sir Oliver withdrew and let his son Raymond speak through the medium. Raymond provides a fascinating sketch of the nature of the adjustments which are made by an individual immediately after "dying." He spoke for half an hour, then returned the use of the medium to Sir Oliver, who went on to speak about his current activities as a spirit on the inner planes.

I was very impressed by the change in personality that occurred when these two fine gentlemen switched places in the session. As might be expected, Sir Oliver was every inch the knight of the realm and university president. It was not difficult to speak with him, but his sense of propriety and dignity was quite evident. As soon as Raymond entered, however, the atmosphere of the sitting changed to a mood of playfulness, quick wit, and youth. This mood remained for the entire time Raymond was speaking, but when he left and Sir Oliver returned, the playfulness disappeared and was replaced by the dignity. They were clearly two different entities.

Sir Oliver mentions his friend Sir Arthur Conan Doyle, the author of the Sherlock Holmes stories, on several occasions in the interview. Sir Arthur was also very much involved in the investigation of psychics and mediums, and did much to make the work of Sir Oliver and others known to the public. Like Sir Oliver, he is a model of the intelligent approach to psychic investigation.

In the last part of the interview, Sir Oliver states that he is still active in helping both psychics and scientists develop their skills. This is a reminder to us that "heaven" is not a distant, faraway location where people rest and sit on their laurels. The spirits of the great geniuses of mankind continue to take an active interest in their work, and are available to help us in our work here on the physical plane. Thus, contact with spirits is not an oddity or wonder—it is to be looked for and cultivated by every intelligent person. The more we become aware of the heaven worlds and their close association with the physical plane, the more we develop a genuinely holistic approach to our own identity, work, and involvement in life. Indeed, we become more aware that we are influenced by cosmic forces of all sorts, and not just our physical environment. That was the lesson that Sir Oliver demonstrated in his physical life—both through his scientific work and his work researching life after death. It is a lesson that is still relevant to us all.

During his career, Sir Oliver wrote many books, both on

scientific and psychic subjects. As far as I know, these books are now out of print, but they are still available in good libraries and used book stores. In addition to the book *Raymond,* which is discussed in the interview, he wrote *Phantom Walls, Evolution and Creation, Ether and Reality, Life and Matter, Man and the Universe,* and *Why I Believe in Personal Immortality.* Many of his writings seek to reconcile the differences between the spiritual and the scientific approaches of life. He saw them as one and the same, and sought to help others see them that way as well.

In the interview, Sir Oliver appears through the mediumship of my good friend, David Kendrick Johnson. Also attending the interview and asking some of the questions was a colleague of mine, whom we will call Dr. Henry Drayton. Dr. Drayton is a medical doctor.

Sir Oliver: Before we get too involved in conversation, I would like to make a few remarks about the present course of psychic investigation, at least from my viewpoint. I'm going to make a flat statement: the people who are engaged in psychic investigation at the present are approaching it in a very pedantic way, and they are going to end up with an empty box. They are looking for the wrong sort of evidence. In the next few years these psychic investigators are going to be caught up short and realize that they have been on the wrong track.

Meanwhile, someone who has been working without fanfare and without seeking fame will emerge on the scene with a sensible program—workable proofs of the reality of psychic phenomena and very workable techniques for developing the latent psychic abilities within people. These proofs will be so utterly demonstrable that there will be no possible argument about them. I'm not going to identify this person, but I can say that the work I am describing will be the impetus of one man in California. You can speculate all you like on who this is; we all have our little suspicions.

Suddenly, all of the *idiot savants,* if I may call them that, are going to discover egg on their faces. These are the psychic investigators who run around with tables and charts and computers and statistics, trying to catalog phenomena. They are all going to be seriously embarrassed. But never mind: they'll be taking with them some of the really bad psychics that they've championed.

Leichtman: Good, I can hardly wait.

Sir Oliver: Certainly the talent, and sometimes even the character, of the psychics these investigators have supported is of very poor quality. As a matter of fact, two psychics who have been sponsored by large organizations are entirely fraudulent. The few insignificant phenomena that they can produce are merely tricks that any good stage magician can pull off with a lot less energy. Their work is not constructive; they don't contribute anything to the human race. In fact, what they do is rather foolish.

The dedicated psychic investigator must learn to discriminate between the frauds and the good psychics—and support the work of the latter. There are many excellent psychics in the United States right now—not thousands of them, by any means, but quite a number—who are quietly going about their work and doing a brilliant job. They are conducting themselves as they should. In return, they are able to make a modest living, but it's a pity they aren't receiving more support. I want to go on record as saying that it's a shame that a good, competent psychic can't expect to make a better living at his or her profession. In many cases a good, compassionate psychic can, in one or two sittings, obtain as much insight into an individual's character as could be produced in four or five years of psychiatric help. This is because an intelligent psychic is able to go directly to the core of a person's problems by clairvoyantly inspecting the content of the client's subconscious and unconscious. It's a pity that kind of work is not more appropriately respected and rewarded.

We used to give Mrs. Leonard [Gladys Osborne Leonard, a British medium who was investigated by Sir Oliver and the Society for Pyschical Research, and who was the primary channel for Sir Oliver's communications with his son Raymond] rather sharp talks about the dabs of money she would charge for very great outputs of energy and time.

I know that you are very concerned about the course of psychic investigation right now. It irritates you that there is so much foolishness accepted as important work, and that there are so many rather childish psychics who are paraded about. If you and other dedicated psychics keep on with what you are doing, in good faith, you will find that your stars are rising. In general, all of the psychics who have been working seriously and quietly thus far will begin to receive the recognition they deserve. Not that they will become famous, because fame is not what they are after, but their work will be seen in a better light.

The foolish must be given the spotlight occasionally so that the rest of society can take a good look at them. It's something like what we are seeing in the recent mass murders out here. The stupidity that has been allowed to grow in society must occasionally be given a strong spotlight so that it can be seen for what it is. In a less dramatic way, that's what's happening in the field of psychic investigation at this time.

I am distressed that so many of the well-meaning scientists who take an interest in psychic investigation do not even use the scientific method! As Mr. Tesla put it, they start with a theory and completely ignore the phenomena. But intelligent people will not be misled by this. They will be able to see that these "scientists" are *not* using the scientific method—or even common sense. You often wonder why it is that some of the least-skilled psychics are the ones who get on the stage with a well-known parapsychologist. Once they are there, all they do is make fools of themselves. But this is all right, because the people in the audience who have the eyes to see will recognize their foolishness—and the foolishness of the "scientist" sponsoring them. It will be these people who will apply the scientific method, even if the scientists themselves do not.

Then, when the star of the competent psychics rises, as it will soon, there will be many people who will greatly appreciate their work—because they will have learned to recognize good quality and important work. I'm not just talking to you now, but to *all* dedicated psychics; anyone who is truly doing good work and is giving of himself and his time will soon find that the world will be more appreciative of his efforts. And there are going to be some very exciting things happening, with all of the good psychics doing some quite important work together.

So don't worry about psychic investigation. Have a little more faith than you do, Doctor, in the scientific method. At the moment, the scientific investigation of psychic phenomena is in uncharted seas, and so it is only natural that the sharks are much more apparent than the good factors. But once the sharks are recognized for what they are, the rest of the fish will be found. The scientific method will ultimately win out.

There is going to be a kind of an Einstein coming out of the scientific community—an Einstein of understanding psychic phenomena. He will produce some formulas that really cannot be refuted. This will do a lot for putting the sharks out of business. I'm sure this will make you dance up and down with glee when it happens.

Leichtman: Well, there certainly is a need for it. Scientists need to investigate the workings of the actual energies and consciousness that are the *cause* of psychic phenomena, rather than the minor physical side effects related to psychic activity.

Sir Oliver: Look at it this way. It is interesting that a large group of scientists is beginning to look at the body of psychic phenomena, whereas in my time *we* were thought to be cranks and crazy. Even in your own immediate experience, you are beginning to see the acceptance of your own work growing a great deal. Both you and David have had your share of figurative rocks thrown at you during the years you've been working, but at least this type of work is beginning to be recognized in the public consciousness. And the scientific community is beginning to take it more seriously, too. Of course, when one is in kindergarten, as the scientists investigating psychic phenomena are at the moment, one tends to be very impressed by all the dazzling "toys" the psychic field provides: floating tables, levitation, teleportation, mind reading, materializations, and so on. So bear in mind that we are talking about an area of investigation in which most scientists have very little experience. Given a little time and maturity, some very serious work is going to be done.

For all of the people out there who have dedicated themselves to this work, there really is hope. Sometimes it does look a little bleak, but psychics themselves often have a tendency to over-emphasize the bleakness. On the basis of my observations during physical life and now, I would say that many psychics are prone to despondency, not so much because of the lack of recognition as much as a lack of *appreciation* of the work they are doing.

Leichtman: I suppose that every individual needs some sort of feedback to reassure him that what he is doing is really accurate and worthwhile, and to support him in his future work. Sadly, competent psychics often do not receive very much of this kind of feedback. Because their work as a rule is not as spectacular as tipping tables or smashing watches, the competent psychics do not receive widespread acclaim.

Sir Oliver: Well, of course, scientists must remember that while a circus is glittering and eye-catching, it is hardly an appropriate arena for scientific investigation. And yet, most scientists are currently approaching psychic research as though it were a circus.

Unfortunately, many people in both the scientific community and the public at large have ulterior motives for seeking out and working with psychics. An ambitious promoter, for example, will find a psychic who is doing spectacular work and will say to himself, "If I can latch onto this psychic, we can make a million dollars together." Of course, what he means is that he will make a million dollars in fees, contracts, publicity, and prestige, while the poor psychic is treated as though he were a trained seal in an animal act. Such people aren't interested in any of the profound implications of the work the psychic is doing: just what they can get out of it. It's an almost criminal attitude.

Many of the scientists investigating psychic phenomena are guilty of having rather self-serving motivations, too. They try to find someone who can be their "pet medium"—someone they can take to conventions and trot out as their own special prize. You've observed this yourself, I'm sure; these mediums are virtually owned by the investigator. They are not well-trained, just showy, misguided, ignorant, and usually self-seeking. These scientists are being very foolish, because once they've latched onto their pet medium they usually don't make much of an effort to test or investigate others. They don't even look around very intelligently; often there are many other people living close by who are far better than their pets.

This type of behavior seriously damages the scientific investigation of psychic phenomena. Once a dedicated psychic has seen a scientist trot out a stupid psychic and make a big fuss over him, he will be very unwilling to volunteer to work with that scientist. My sympathies are with the good psychic in such cases. It would be so easy for scientists to find better quality psychics to work with, and that in turn would guarantee higher quality results from their experiments. They could at least look at the bulletin boards of the local metaphysical bookstores and visit the psychics who post notices there. If they intelligently interviewed several psychics, letting them explain their work without a barrage of criticism, they would be able to find the serious, thoughtful people in the field. Unfortunately, most of the psychic investigators of today are a bit on the lazy side: they are so busy trying to make a name for themselves that they don't bother to carefully seek out and find the very best psychics.

Another pitfall in the current investigation of psychic phenomena is the age-old problem of scientific and academic endeavor: doing work to please one's colleagues instead of promoting true scientific inquiry. As a result, many psychic investigators are trying harder to gain recognition from their professional colleagues than to make any type of valid discovery. This is one reason why so many of them pirate each other's experiments and ideas.

It's time for parapsychologists of the stature of Dr. Rhine [Dr. J.B. Rhine, the acknowledged "dean" of American psychic investigators] to publicly state that we have already proved that ESP exists. The need for further testing of its validity has passed. Now it's time to find out what makes psychic phenomena work.

Leichtman: Yes. Let me ask you this question: should investigators of psychics and psychic phenomena have psychic ability themselves?

Sir Oliver: There are several qualifications a good psychic investigator should have. Arthur [Sir Arthur Conan Doyle] and I and the others we worked with found, after a time, that we got much better results when we relied on our intuition, yes. As you know, intuition is one of the psychic gifts that everyone in the world has. We also found that it is important to go into an investigation with an open mind, maintaining an intelligent degree of skepticism. But perhaps most important, we learned the value of treating the psychics and mediums we investigated with simple, common courtesy. Both of us were very interested in finding genuine psychics and mediums. We discovered that if we approached a psychic with courtesy, kindness, and a sense that he or she was an intelligent participant in the human race, then the results of our experiments were much better. Many modern scientists tend to think of psychics as being slightly crazy and treat them in very condescending ways. That kind of an attitude guarantees that the test results will be of a very low quality—after all, a psychic is a human being, and if one is treated in a condescending manner, one is offended. A psychic cannot do his best when he is not being treated with courtesy and respect by the investigator.

The good investigator will be able to maintain his skepticism without being discourteous or insulting to the psychic. It's good to be skeptical; in fact, any intelligent psychic enjoys working with skeptics as long as they are fair-minded and genuinely interested in discovering the truth. A skeptic is going to cause the intelligent psychic to examine what he or she is doing a little more closely—to give it closer scrutiny. That's good for anyone.

Leichtman: Of course, there is always the consideration that in order to produce valid and accurate psychic work, the experiment has to be something that is useful and fulfilling—not just some totally meaningless and purposeless card-turning trick. I imagine that investigations often come to naught because the psychics are asked to deal with such utter trivia, that is important to absolutely no one.

Sir Oliver: The card experiments did have the virtue of demonstrating, in the beginning, that there was *something* happening that could be measured and verified. I agree with you wholeheartedly that a gift, whether it is a gift to play the piano or a gift to be psychic, has to have some kind of constructive grounding in the physical plane. It has to be developed into something more than just being a parlor game or the subject of idle curiosity. The gift becomes useless unless it is given meaning in the context of real life needs. I would like to see more people approach psychic investigation from the standpoint of harnessing psychic talents for making useful contributions to mankind, rather than just gathering statistics. That's a bit silly.

Leichtman: Abilities are to be used, not just measured.

Sir Oliver: Yes. Unfortunately, science has lost sight of its dynamic purpose. We seem to live in an age in which the worker doesn't want to have to work. In the field of science, many scientists have allowed themselves to degenerate into simple statisticians. That's a prostitution of the scientific method. It is *not* science—it's the great whore of science, if I may put it that way. I mean to put it very strongly.

In investigating psychic phenomena, what counts is the validity, the applicability, and the genuineness of what occurs—how it helps explain physical life and the nature of consciousness.

Leichtman: Right. Let me ask you another question.

Sir Oliver: If you don't mind, let me ask you one of my own. What did you think of *Raymond?* I'm curious, not so much what you thought of the content as what you thought of the style. We can talk about the content if you want, too. I'm mostly interested in the modern reaction to the style of writing I used.

Leichtman: It was difficult to read in many places; at times the sentence structure was a bit pedantic. In the third part, perhaps it would have been possible to present some of your images and ideas a bit more lucidly. And in the second part, where you review the evidence and the sittings with the mediums, it got a little tedious to keep track of exactly what was going on. Of course, I'm aware that you were often condensing two hours of a sitting within two pages, but it was a bit disjointed.

Sir Oliver: Those remarks are very interesting, because I had hoped that the book would continue to be read a bit longer than it was. I was trying to put down some controversial facts in a way that was readable and somewhat entertaining. I was experimenting with what was actually a new area of writing.

Leichtman: I'm very happy you put in the first part [which dealt with Raymond's childhood and experiences in the armed forces before he was killed]. Some people might have objected to that. But the first part has to do with—

Sir Oliver [interrupting, by standing up]: Excuse me. I'm going to do something here. *[He proceeded to open a window.]*

Leichtman: I was going to do that myself—it's getting a bit stuffy.

Sir Oliver: Getting up and opening a window is not even a "dramatic first" for a seance. Using the medium's body to do things like this is rather more common than is ordinarily supposed.

Leichtman: For the record, Sir Oliver has taken David's body to the window and has opened it, and is now returning to sit down.

Sir Oliver: One thing about it, David knows the house

rather well in the dark [the session was being conducted in the evening], so it makes it a little easier for me. I would hate to try that experiment in your house. That might be disastrous.

Leichtman: In my house, I don't have any windows *to* open. I have a balcony you could fall off of, but that's about all that could happen. *[Laughter.]*

Sir Oliver: One of the good things about doing a trance through David is that his body is very usable. And you, too: Mr. Tesla reports how easy it was to work through you. There is no reason why any of us could not get up and open windows and do other things. As a matter of fact, it is quite commonly done. It's only the old school spiritualist types that make the claim, "Oh, you don't know how much effort it is just to sit here and go into a trance."

Frankly, at this stage I am sorry that I was connected to the work that gave impetus to spiritualism. At the time I was involved, spiritualism was not what it is now. Even in England, spiritualism is still a good deal better than what it has become in America. As a matter of fact, there have always been thoughtful, intelligent people involved in the Spiritualist movement in England and not nearly so many uneducated people.

As I look at spiritualism in the United States, I find that it has become antiquated, when compared with everything else that is going on in the psychic area. Spiritualism is not going to last much longer in this century, at least in the United States, unless some young leadership that is willing to revise its attitudes and be more dynamic comes along and is accepted by the Spiritualist Church.

The part of American spiritualism that is so very bad is also a problem in England. Fraud is too frequently condoned on the rationalization that the psychic is merely having an "off day" and is therefore entitled to fake it. This is not acceptable! To paraphrase Gertrude Stein, "a lie is a lie is a lie." There can be no excuse for cheating.

Leichtman: I agree.

Sir Oliver: I also think that when American spiritualism

comes up with the statement that there can be Christian Spiritualists and non-Christian Spiritualists, it is ringing its own death knell.

Leichtman: It was abundantly clear to me from reading *Raymond* that you found great solace and hope in the fact that certain psychic phenomena and the survival of the human personality after death could be validated—and that these proofs must be part of the evidence indicating a creative intelligence that governs all. And yet, it is this type of realization that seems to be so sadly ignored in modern spiritualism—as well as most modern investigations of psychic phenomena.

Sir Oliver: I am amazed that anyone can get interested in psychic phenomena, develop a little talent, but not realize that there is a god behind all of life. Yet many people seem to be able to do trance work, talk to discarnate spirits, perform magic, or do any of the wide spectrum of psychic feats and not realize that there is a god behind it all, somewhere. I just don't understand it, because the reality of God is so blatantly apparent to me. It was even during my physical lifetime.

Leichtman: I must add one extra comment that I didn't get finished when you got up to open the window. I do think the first part of your book, in which you incorporated the letters of Raymond, was a vital part—it gave a living presence to the book. It must have done a great deal to help a number of people in a time of great personal tragedy—people who had also lost someone very dear to them. The book was far more than just an academic or a scientific pursuit.

Sir Oliver: The book was extremely personal. Bishop Pike [Episcopal bishop James Pike, author of *The Other Side*] and I have discussed on a number of occasions the fact that the intensely personal search is really the only valid one. These discussions have been since he came over to the inner planes, of course, not while he was still in physical life.

I sometimes think that the investigation of psychic phenomena is in the wrong hands when left to the scientific community. It should be more in the hands of philosophy or religion—one of the two. But then, the way religion is organized, particularly Christianity, it would be *very* difficult to incorporate any effective psychic investigation into its body of thinking at this time.

I mentioned that I was amazed that anyone could explore the psychic world without realizing that God is behind it all. Well, I'm just as amazed that Christians, who are taught that the soul is immortal and never really dies, are nonetheless so utterly offended by evidence that it is possible to converse with someone who is "dead." Imagine it—someone who has been dead as long as I have been can still carry on an intelligent conversation! *[Laughter.]* If the Christians who don't believe in the possibility of talking with spirits would only think about it for a minute, they would see the contradictions in their beliefs. Why should I or any other "dead" person be far away and unreachable? It doesn't make any sense.

This is another personal reason why I followed the line of investigation that I did. I think you and David share this same perspective with me.

Leichtman: Definitely. What role do you think psychics and mediums should play in terms of communicating with those who have passed out of the physical plane? Is it still important to accumulate evidence that such communication is possible, or do you consider that the fact of the survival of the human personality has now been established beyond any reasonable doubt?

Sir Oliver: Oh, I think it's been proven. Of course, there will always be magazine writers and narrow-minded "scientists" saying that all of it is fraudulent and fake. But any sensible, open-minded person looking at the material that is available will realize that the survival of the personality after death has been established beyond the shadow of a doubt. There have been quite a number of intensely personal accounts that strongly demonstrate survival—Bishop Pike's book is an interesting example in the present time. And there have been a number of other serious investigations that have produced magnificent results—such as the work done by the Whites [Stewart and Betty White]. Their investigations produced ideas and insights that went far beyond their ordinary thinking; it cannot be reasonably argued that the medium pulled these ideas out of someone else's head, or something the medium read. The accounts left by these and other people should be enough to demonstrate that it is possible to communicate intelligently with spirits. It's also time for the people who frequent mediums to realize that the purpose of communicating with spirits is far more serious than just sitting down every Thursday at two o'clock at your "friendly neighborhood Madame Zenobia's" to exchange gossip with spooks. I'm not saying that it's not valuable for the recently bereaved to be able to contact their deceased friend or relative long enough to demonstrate to them that this person is fine and not that terribly far away. That can be very helpful. But endless sittings with Aunt Minnie are really childish—and they interfere with what Aunt Minnie should be accomplishing over here. We're all very busy here in heaven, you know. We have things to do. Life doesn't stop just because you no longer have physical fingers and toes.

Did you hear me applauding when you were interviewing Madame Blavatsky, and she was giving her definition of what it is to be a spirit? I was, because that's essentially my outlook, too. We are still involved and still care about what is happening in physical life. People might wonder why I come back and talk through a medium as I am doing now. It is because I'm still involved in life and curious about life. My perspectives are slightly different than they were, and I'm concerned with a fuller view of life than most physical people are, but that doesn't mean that I'm going to exclude the physical.

Leichtman: To follow up on that: from your current perspective, how would you like to see mediumship applied in this day and age?

Sir Oliver [chuckling]: I think you're fishing for an obvious answer, but I'll give it to you anyway. Just like everyone else participating in this project, I feel that the best use of mediumship is the kind of work we're doing here. Some other examples of the ideal use of mediumship would be the work done by the Whites, by Darby and Joan in preparing *Our Unseen Guest,* and even by Jane Roberts in bringing through the Seth material—although I know you'll grit your teeth about that last one, Doctor. These are all relatively important contributions to human thought. Obviously, it's much more worthwhile to talk about the important topics of human life, as we are this evening, than to just tell someone personal information. Besides which, David can give out personal information and advice to people who need it without going into a trance. It would be a waste of my time to call me in to do that kind of psychic work. I believe Dr. Kammutt [Dr. Kammutt is an inner planes teacher who works with both Dr. Leichtman and Mr. Johnson] has been rather outspoken at times in stating, "David could do this himself without calling me in." In point of fact, much of the work done by many mediums is something that could be better done by any competent psychic. It would be preferable to reserve mediumship for worthwhile conversations with intelligent spirits.

It's a real mistake to become too dependent on spooks. I'm endlessly amused, for example, by the people who try to explain Christ's greatness by saying that His spirit guides did it all for Him. That's just an intellectual flip-flop, and rather childish. It's too bad that some people are so self-centered and feel so threatened by the greatness in others that they have to devise some way to pull the great down to their own level. One thing that humanity has to learn is to embrace its own greatness— its greatness collectively and the greatness of each individual human being. Each person must eventually embrace his own genius, because we each do have the potential of genius. Sometimes it's buried under tons and tons of turnips, but nevertheless the capacity for greatness is there, because the Maker of all things imbued the whole spectrum of life with genius. It is really a religious duty for a human being to embrace that potential and live by it.

One part of genius, of course, is psychic awareness.

Leichtman: I agree that developing genius is a religious duty. I know many other people, however, who would claim that our religious duty is to be humble and meek. Often, that inhibits the development of their talents.

Sir Oliver: If by being humble and meek you mean that we need to be charitable and considerate, then yes. We need to participate with the rest of the world. Certainly, we must never glorify self-centeredness. But the so-called Christian ethic sometimes loses sight of an important idea: our body and life are God-given. Therefore, there is an element of greatness within them. And we must care for and nourish this divinity. Our individual well-being and growth is part of our responsibility. It has been said and often quoted that a person comes into physical life to be spiritual and to help uplift as many people and as much of life as possible. But the next sentence is usually omitted: the individual has that same responsibility to himself as well! It is part of the duty of every human being to nurture his own greatness in addition to nurturing the greatness in others. Just by observing the animal and plant kingdom, the inquiring person ought to be able to see that all of creation is designed to take care of its needs, its health, and its well-being. Then it shares anything that's left over. This is not selfishness. Creation is designed to take care of itself.

Leichtman: Yes, it has to be.

Sir Oliver: Let me change the subject. I want to talk about the astral. Back when I was working, we were a little more involved with astral phenomena than you are at the present time. And the astral plane was a good deal cleaner than it is at present. It didn't have all the clutter that it has now.

Leichtman: Why is that so?

Sir Oliver: There are several reasons. First of all, the general public is less restrained and more emotional than in my time. There is more emphasis on expressing feelings, especially uncomfortable ones. All of this happens on the astral plane, and as you know, physical people contribute to the quality of the astral every bit as much as those who have passed on. So, the emotionalism in the world today has become a serious problem.

Another contributing factor is the tremendous growth of public interest in psychic phenomena and magical practices. In my day, the *masses* were not that interested in psychic matters—they were afraid of them and stayed away from them. Even the people who did investigate the psychic did so with great caution and prudence. Nowadays, many people embrace these phenomena almost too readily—without a healthy respect for their complexity or potential danger. As a result, some groups actually lust for psychic experiences, without regard for their type or quality. This psychic lust energizes all kinds of low quality astral phenomena.

Leichtman: These seem like very serious problems.

Sir Oliver: Not as disturbing as another of your modern phenomena—the growth of mass communications such as television, radio, movies, and the like. There are many good qualities in all of these media, but the public seems to be attracted most strongly to the elements that are shocking, morbid, titillating, and disgusting. I don't want to sound like a prude, but I am greatly aware that the typical public reaction to these forms of entertainment—and sometimes even to what is presented as news—has been perhaps the greatest contributing factor to astral clutter.

Leichtman: Are you implying that there were less shocking, morbid, titillating, and disgusting events in your day than now, or just less awareness of them?

Sir Oliver: Neither. The key factor is the public's *passive reaction* to what it sees and hears. People are almost automatically reflecting into the astral the ugliness around them. This can be seen, for example, in the recent fascination with disaster movies. As people watch these films, their emotional reactions become mirrors of the shocking and destructive themes and images in the movies themselves. As these reactions accumulate collectively, they become virtually a tidal wave of disaster on the astral plane.

Leichtman: Do you mean that much of the clutter on the astral plane is really the reflection of people's unenlightened reactions to physical events?

Sir Oliver: Yes, and your mass media has made awareness of tragic and shocking events, whether real or fictional, much more accessible to the general public than in my day.

Even something as commonplace as the popular music of today adds to the clutter of the astral. When you have millions of people feeding themselves on modern jazz and rock and roll, which are more violence and lust set to sound than they are genuine music, the astral problem that results influences not only the perceptions of psychics, but the feelings of ordinary people as well.

If you don't mind, let me tell you a little story; then, I want to let David up. He's getting thirsty, and if we're going to go on talking, we're going to have to give him some water. This is a story about astral sight. There are a number of amusing incidents that happened that never got written down. Arthur was a bit reticent to talk about some of these occurrences.

We had one very delightful session with Mrs. Leonard. We had gone to a ruined abbey during an afternoon, because Mrs. Leonard could work in the daylight, too. She was happy to, in fact. She loved nature very much and did not have many opportunities for an outing, so we made kind of a picnic out of it. All of a sudden we were surrounded by little fairies. Everyone was secretly delighted, of course, but because we were adults, we tried to be appropriately shocked. *[Laughter.]* It was a wonderful experience.

Of course, this kind of experience would be less likely to occur nowadays, due to the increase in astral clutter we've just been talking about. Thought-forms of negativity and discordance offend the "little people."

Nevertheless, there will soon be an increase in the old "fairy faith," as it was called in England, because more humans will be developing clairvoyant awareness and will be able to see them in the realms they have retreated to. Some of these people will also become more concerned about creating the kind of psychic and emotional atmosphere that will attract the fairies to them. In fact, there's one peeking over your shoulder right now. A little, tiny thing. They are lovely. Of course, all of God's creation is very lovely, and that includes the invisible portions of creation, too. Well, I'm going to step out for a few minutes.

Leichtman: Okay.

[During the break, various forms of psychokinesis (see glossary) were discussed, including "table tipping," a method of psychic communication that was used on occasion by Sir Oliver and his family. After David went back into trance, Sir Oliver picked up the thread of the informal conversation.]

Sir Oliver: A short discourse on the mechanics of table tipping would be appropriate at this point in the conversation—and it's going to be a short one because that's all it's really worth. There are a number of phenomena connected with communicating with the dead that have been frequently studied by psychic investigators to see if they are contrived or real—phenomena such as table tipping or trumpets floating in mid air. Sometimes these phenomena are faked, but when they are genuine they really aren't quite as mysterious as they seem to be. Frankly, they are so easy to manifest that I wonder why scientists haven't caught on yet. The force that makes a table tip psychically, for example, is just the strong force of the psychic energy of the physical people involved. Often, it's a strong negative emotion such as anger or hostility that is turned into the force that moves the table, or whatever the physical object is. That's all table tipping amounts to. The physical plane is already so victimized by anger or hostility that a responsible psychic investigator should think twice before encouraging the use of these techniques for physical mediumship. Why set loose on earth strong negative forces that we already have a plethora of?

Leichtman: Are you including all types of psychokinesis in making that statement?

Sir Oliver: Not all types of psychokinesis, because there

are exceptions. But I am including table tipping, the use of the Ouija board, and other common forms of psychokinesis. All of these are basically poltergeist phenomena, in that the movement is caused by some kind of repressed emotion. They are astral phenomena. The energy comes from the physical people sitting in the room, not from another intelligence or another source.

Leichtman: I believe in your book *Raymond,* you speculated that the motive force behind table tipping was being supplied by the physical people present, and that the spirit intelligence merely utilized the energy made available to him.

Sir Oliver: Yes. But I'd like to encourage people to stop thinking so much about the glittering uses of telekinesis in seances and realize that it has practical uses, such as the ones you mentioned during the break. I eavesdrop a lot, you know. *[Laughter.]* For the record, you were talking about going off on a trip and realizing that you had left something you needed on the kitchen table at home. You mentioned that a practical use of psychokinesis would be the ability to transfer psychically the needed item from the kitchen table to your hotel room, so that it would be there when you arrived. I agree that that kind of telekinesis would be very useful. And almost anyone can do it. It's just a question of how well you are attuned to yourself and your own power.

Leichtman: Almost *anyone* can do that?

Sir Oliver: Certainly. It has been recognized by European scientists—who are prone to be, I dare say, a little more romantic—that the human organism has available to it at least one hundred times more power and awareness of its surroundings than it needs to survive. Unfortunately, most people in the world today use only a small fraction of that power and awareness. An intelligently aware person may use twenty-five percent of his capacity, but most people use only about five percent. You can't expect much out of the type of person who shuts out facts or shuts out the invisible, psychic dimensions of life.

Leichtman: Getting back to the subject of communicating

between a spirit intelligence and a physical person, would you comment on the various types of communication that could be used—their usefulness and validity? For instance, in contacting your son Raymond after his death, you used trance mediumship, the direct clairvoyance of a conscious psychic, automatic writing, the Ouija board, table tipping, and some others.

Sir Oliver: The technique itself is not important; what is important is the consciousness of the sitter [the psychic or medium making the contact]. If, for example, you were trying to communicate through a trance medium who had the consciousness of a child, you would get very childish information. Indeed, you might even get a very childish spirit or an undeveloped elemental. You might contact a spirit who was never very intelligent or aware—one that never grew up much while alive physically, and hasn't changed much since passing over, either. David ran into a case like this once at a Spiritualist church. The medium went into trance and a Calvinist pastor came through who hadn't changed his views in a hundred years. David thought it was terribly comical that he had to go to a Spiritualist church to hear a Calvinist minister!

[Much laughter.]

On the other hand, you could go to another trance medium and have a very intelligent conversation with a knowledgeable spirit. That would be very constructive. So it depends on the consciousness of the medium and the spirit you're talking to, not just the process.

If you try to use automatic writing and don't happen to be mediumistic, you'll probably just contact your own subconscious and not a spirit. As you know, automatic writing is a tool that can be an excellent way of intelligently tapping your own subconscious, if it's used under some kind of expert supervision. But a mediumistic, intelligent person who has taken the trouble to investigate himself and train himself carefully can use automatic writing and make contact with helpful and intelligent spirits. In such a case, you're likely to receive important and pertinent communications. I know that both you and David frequently use a variety of automatic writing in just that way.

I dislike the Ouija board. Count Hamon [Cheiro] wants to talk about Ouija boards in his interview, I believe. I'll just say that the Ouija board has no safeguards; one is never sure what he is getting. It could be a spirit or just one's own subconscious. Let me tell you about one of our own experiences. We ran into several people who were using Ouija boards and who seemed to be receiving impressive communications. That was the rumor, anyway. But once the messages were carefully examined at the end of the sitting, they were found to be gibberish. Even for the best of psychics, the Ouija board is really not a reliable tool. In fact, a person with a high quality of consciousness and contact with higher intelligences on the inner planes cannot use a Ouija board at all, because the spirits simply refuse to work through that means. The Ouija board is really a child's toy that a highly-evolved person would have outgrown. A very dangerous child's toy, I might add. The Food and Drug Administration ought to add a little label to Ouija boards warning that they are injurious to your health—just like they have on cigarette labels. Ouija boards are almost as injurious to most people. Of course, there are exceptions: some people do manage to get profound messages through them, but they are very few and far between.

Leichtman: Wouldn't these people rather quickly switch to using another, more efficient method of communicating?

Sir Oliver: Yes, they would find another, easier method.

Leichtman: Direct clairaudience, automatic writing, or something. While you're on this subject, would you care to expound a bit more on exactly what the dangers of a Ouija board are? Are you referring to dangers of confusion or possession—or what?

Sir Oliver: That can best be answered by listing the kinds of contacts that can be made on a Ouija board. Let's take a hypothetical case, where the sitter [the person using the Ouija board] is an average American who is slightly psychic, is fairly intelligent, has a middle-class background, perhaps has several years of college, but who has never resolved the emotional conflicts of his or her life—like most Americans. The first contact this hypothetical person would make would be with the wish fulfillment department of the subconscious. The kind of message that would come through would be: "Oh, my dear saint, the world just doesn't understand you or appreciate you. Everyone you meet on the street should genuflect because you're such a great, important spirit." *[Laughter.]*

The danger here is self-deception. As someone else has said during this project, it's just a case of the subconscious being a willing servant of the conscious mind—putting on a sheet and playing spook.

The next type of contact that can be made with a Ouija board is likely to be a malicious elemental—because there are no controls on Ouija boards. Elementals don't start out being malicious, but they can acquire that characteristic if they come into contact at some point with malicious people. For instance, some of them are attracted to satanic covens and are taught some very unpleasant manners. This type of elemental often comes through Ouija boards.

Bear in mind that there are elementals of differing quality: some are good, some are bad, some are neutral. It depends on the type of people they have been drawn to in their careers.

Leichtman: Would you define "elemental" for the record?

Sir Oliver: An elemental is a living entity, but not a human being. Elementals are part of a separate line of evolution in the cosmos. Elementals don't have physical bodies; most of the ones a person might contact through a Ouija board exist in the lower reaches of the astral plane.

Leichtman: Don't some elementals exist in the etheric?

Sir Oliver: Oh, yes, but the elementals most commonly encountered by psychics are on the lower astral. These are the beings that gave rise to the common European ideas about fairies, elves, gnomes, dryads, and the like. And as you know from the European legends, not all of these creatures are always beneficial to man. There are all sorts of stories in the folklore of Europe and other places about the little people being very malicious unless they have been brought up to be good. These stories should give you an idea of the kind of entity you can come up against in playing with the Ouija board. An elemental is likely to start giving orders: "You must divorce your spouse." "You must quit your job." That sort of thing. Very destructive advice can come from contact with these elementals. Of course, it's also possible to run into a nice elemental, but it's unlikely that you'd contact one of them through a Ouija board, because their consciousness is already beyond the Ouija board. A nice elemental knows that it can come directly and be helpful, and be rewarded in kind. And the elementals are rewarded for their help, because the universe is based on the principle of exchange. "Exchange" is one of the meanings of the word "karma," incidentally.

Just to set the record straight: a good elemental can come without a Ouija board and make itself helpful. A bad elemental is something like an egocentric person that must always have a platform to speak from; therefore, it is greatly attracted to Ouija boards. They are ideal platforms for these elementals. They think of it as a game. They are not really immoral, just amoral. After all, what can you expect?

Some of these amoral elementals are very cunning types that can take the sitter's thought-form of a period of history or a figure in history and animate it. Then the following kind of message would be spelled out on the Ouija board: "This is Queen Elizabeth I. I have something important to say to you." Of course, it isn't—it's just an elemental animating the human's thoughts.

Leichtman: How can one know it's an elemental and not the real queen?

Sir Oliver: Usually it's very obvious—and frequently comical, too. We are using a hypothetical case of an average American, right?

Leichtman: Yes.

Sir Oliver: And is it fair to say that the average American has a pretty bad grasp of history?

Leichtman: Yes.

Sir Oliver: In fact, the average American would probably imagine Queen Elizabeth in one of the ways that she's been portrayed by movie actresses—whichever version made the strongest impression. If our average American is a woman with the standard American female consciousness (I don't want to sound like a male chauvinist, because I am not), she probably thinks of Queen Elizabeth mostly in terms of the gowns and the other trappings in the film. In other words, her thoughtform of Queen Elizabeth is defined more by the color of the period than by historical fact. She emphasizes the aspects of Queen Elizabeth that have been so heavily emphasized in the films: the idea that this is a woman of great power who somehow had men entwined around her little finger. I'm using Queen Elizabeth as an example, but I could use any historical figure.

Now, most average Americans are a bit disappointed by their intimate lives and fantasize quite freely. And so she may add a few characteristics to the queen that aren't even in the movies. What we end up with is a cardboard image of a historical figure that is not fleshed out much beyond the gown and the hairdo, with a few obvious and imagined traits. She may very well have a more glamorous face than Queen Elizabeth ever possessed, because most of the actresses who have played Queen Elizabeth have been better looking than the queen, even on their worst days. She may hit me for saying that. *[Laughter.]* No, she won't—she's too kind. She's a good egg, when all is said and done.

Bear in mind that there is no substance of the real queen in this cardboard thought-image. The only substance that survives is the few facts that accidentally found their way into the films. Perhaps I'm overstating this a trifle; I do realize that there has recently been a more accurate portrayal of Queen Elizabeth in a television series, but even this is not complete. The thought-form of the queen is not a full being. It cannot be aware of itself. It is not the shade of Queen Elizabeth. It is a cardboard thought image of the queen in the mind of our hypothetical sitter. And as this person is using the Ouija board, she contacts an elemental who then draws forth from the sitter's mind this thought-form and animates it as though it were a puppet. Quite literally, the elemental wears the thought-form of Queen Elizabeth as if it were a dress. And the message that comes through on the Ouija board says: "Hello. This is Queen Elizabeth. You are my long-lost descendent. You are really the crowned princess of England, and no one has ever recognized this in you." *[Much laughing and guffawing.]*

Because the average American has such a bad grasp of history, these messages always sound a bit ridiculous. After all, Elizabeth never married. She didn't have any descendents! But the elemental doesn't know that—and neither does the sitter.

Any number of animated thought-forms can be produced in this way. We can have an animation of Amenhotep IV, if you will, or of the Colossus of Rhodes [Apollo], if one has enough imagination for that. *[Chuckling.]* It's based on the sitter's imagination.

Leichtman: Even Scarlett O'Hara?

Sir Oliver: Even Scarlett O'Hara, who never did exist. As a matter of fact, I think that you and David had an experience with a lady who "contacted" Scarlett O'Hara on a Ouija board. *Gone With the Wind* is set in a period that holds a great deal of romance for many Americans: a period when life was gentler and a bit more structured, more refined, and more glamorous. Or at least that's the cardboard image that many Americans have of that period; it's not really based on a complete grasp of American history. Rather, it's the cardboard fantasy of the past that is promoted by the public school system in America, which unfortunately never quite gives its students an idea about what history really deals with.

Scarlett O'Hara may be a composite of people that Margaret Mitchell knew in life, but she's a fictional character, not a human being. Nonetheless, she seems to be coming through Ouija boards with increasing regularity! Every time the film is released again, several people get messages from her.

Leichtman: She's a bit passé now. I suppose nowadays the elementals should start moving on to characters like Spock on *Star Trek* or James Bond. *[Laughter.]*

Sir Oliver: Whether the person using the Ouija board contacts his own subconscious or an elemental, the experiences resulting from these contacts are often very destructive. There can be a lot of self deception and malicious advice. People have been known to get divorced because their Ouija boards told them to, for example. That's very unfortunate but not uncommon. One of the problems is that most people who use Ouija boards have not yet delved deeply into the literature of the occult, which would warn them to use better means. They are just beginning to play at being psychic.

For the record, however, let's assume that our hypothetical sitter is able to recognize the undesirable contacts I've just listed for what they are, and has not been deceived by them. If she has that much discernment, there are some other contacts that she might make that would be more beneficial. For example, she might contact the genuine spirit of a deceased relative, who would be able to make an intelligent statement about what it means to be alive and well on the inner planes. That type of contact, however, would tend to be rather brief. The other type of contact she could make would be with a teacher on the inner planes. Very soon afterwards, however, she would begin to realize that the quality of the consciousness of this teacher is entirely different from anything else she had previously contacted. And the teacher would also very quickly insist that the sitter stop using the Ouija board and learn to communicate in a more efficient way.

What I am saying in all of these comments is that it doesn't really matter what technique one uses in making contact with the inner planes; what does count is the validity and the applicability of the insight that is received. What counts is the genuineness of the information that comes through, in terms of consciousness and its usefulness in physical life—not the means by which it comes.

Leichtman: Right. Let me ask a couple of other questions on this subject. Is it possible to contact not only your subconscious through the Ouija board but also something of your own spirit, at least temporarily?

Sir Oliver: Oh, yes—although it's unlikely to happen with the Ouija board. Again, Count Hamon is suggesting that you make a note to ask him this question as well. It is quite possible to contact your own inner being psychically, yes. This is what happens when one starts to remember valid experiences from past lives. These "memories" are part of the inner being— part of the subconscious of the inner being, I suppose, except that things are reversed at that level, as you know.

There are several valid techniques that can help teach you about the contents of your mind and the mechanics of it. A good system for reading cards, for example, can be quite valid in helping one discover what's in his head and how it all works.

Leichtman: I don't want to take material from Count Hamon, but what's happening when a person begins to get psychic messages that are very derogatory or malicious in character? Is this the unredeemed aspect of his or her subconscious that is being contacted?

Sir Oliver: Not always. There are malicious beings who are quite willing to give malicious messages, and I'm not talking about elementals now. I'm talking about human beings. As it has been said in the past, just because someone has passed over does not necessarily make him wise. And so it is very important to remember the Biblical injunction, "By their acts ye shall know them."

Leichtman: Would you care to comment on the varieties of automatic writing?

Sir Oliver: In my day, automatic writing was done somewhat differently than it is now; it was a very tedious method. There were two techniques used. In one of them, a pencil was inserted into a planchette like the type used in a Ouija board; the movements of the planchette produced the automatic writing. This method had all of the dangers of the Ouija board, of course. In the other method, the pencil was propped in the hand in this way. *Sir Oliver uses David's hand to demonstrate.*] It's hard to describe exactly—the pencil was propped on the outside of the hand. It was not held in the ordinary way. This particular technique was very tedious and took a great deal of effort to make it work.

An easier version of automatic writing is to grip the pencil in the ordinary way, positioning the hand on a piece of paper, and then relaxing the mind, trying to "fall into" yourself. That's the best way I can describe it. You let the hand write down ideas without directing it from your conscious mind. This type of automatic writing is sometimes used in psychotherapy, is it not?

Leichtman: Yes, it is.

Sir Oliver: Of course, the quality of the messages that come through automatic writing depends on the development of the individual, his grasp of how his own mind works, and his ability to transcend himself, so to speak. A well-trained person can receive dictation clairaudiently which is directly translated to the hand, almost without passing through the mind. It's just an automatic function of the subconscious and the body. But as before, the quality of the content is more important than the method by which it is received.

Automatic writing has become a good deal easier since the advent of the typewriter. It's just a question of turning your hands over to someone on the inner planes to type with. David is telling me about a time when his hands were typing a message and he was whirling around and laughing on the floor. I find it hard to visualize that, but it was true automatic typing.

These new methods of automatic writing are far more efficient than some of the older ways. The ways we did it were more for show than for anything else.

Leichtman: Very good. I think that covers the various

forms of communicating between spirits and physical people, unless you wish to add anything.

Sir Oliver: Since you ask, I would like to make some remarks about Mrs. Leonard. I don't wish to blow the lady's horn, but I do have something to say that will be pertinent here. Of all the mediums we consulted, I found Mrs. Leonard to be the easiest to work with. I would attribute this to Mrs. Leonard's bearing and her level of consciousness. She was so exquisitely honest that you were sure that anything she did would be honest—or she wouldn't do it. She was a very intelligent lady who had beautiful manners and was very considerate of others. She cared a great deal about others and gave of herself endlessly.

Now, my reason for saying all this is to emphasize the fact that a medium's intelligence, basic character, and bearing are very good gauges to use in evaluating the quality of the work he or she can do. Anyone seeking out a medium to consult should take a look at how that person deals with life in general, because that has a very important relationship to the type of communication they can bring through. Mrs. Leonard was honest and so the beings on the inner planes who were attracted to work with her were also honest, and gentle. The same rule applies to you and David and anyone else: you tend to attract to you beings from the inner planes that treat the world in much the same way that you treat the world. This is because the universe operates on the Law of Exchange.

Leichtman: While you are talking about Mrs. Leonard, wasn't her control a spirit named Feda?

Sir Oliver: Yes. She had more than one control, of course, but not all of them were written about.

Leichtman: I was rather amazed that Feda had, frankly, such a childish mentality, and wondered if there might not have been someone better for Mrs. Leonard to use.

Sir Oliver: In a way, David's friend Rosebud [a spirit who acts as David's mistress of ceremonies] has a childish mentality, but you know from talking to her that she really isn't childish. Feda—

Leichtman: Feda lacked clarity of expression.

Sir Oliver: Yes, she did have some problems in that department. I'm at a loss as to how to describe to you the difference in sitting and listening to Feda and the way her comments appear in print. There was a genuineness about the contact that just can't be put into print. It's hard to explain: Feda was a child, but she wasn't childish. She is one of those beings who has seldom had an incarnation where she lived much beyond the age of eight or nine. So her consciousness on the inner planes has many of the characteristics of a child. She did remarkably well as a control, if you consider this factor.

Now, Rosie is girlish; I suppose "childlike" is a better description of her. She's standing here frowning at me, making sure I get this right. *[Laughter.]* Rosie did live to be a mature woman; she's simply a being who is full of the joy of living. Feda partook of this spirit, too. It's hard to describe because she had certain qualities that just don't have names in English. And any English that you might try to use to describe what made her an effective control would be a distortion, really.

Leichtman: David and I have always used clarity of expression as one of the major criteria for evaluating a psychic and a psychic's work. Sometimes that seemed notably missing in Feda.

Sir Oliver: Of course, at that time we were delighted just to find *anything* that seemed coherent and genuine. Please remember that we looked around a great deal in choosing mediums, and we found many people who had controls that were even less coherent. Part of the problem in this regard was Mrs. Leonard herself; she was a little bit terrified of her own work. No, "terrified" is not the right word; she held the work in a certain degree of awe that sometimes interfered with communication. She frankly admitted this problem to me—she did not completely understand all of the implications of what she was doing.

Sometimes the mechanics of this kind of communication go awry, and you must remember that not all of the mediums in my day had the opportunity to be as well trained as you and David and some other modern people are. That can cause a breakdown in the mechanics of a given communication. I remember the first contact you had with Bishop Leadbeater on this project—for a time at the beginning, he spoke with a certain circuitous, concentric sentence structure that will have to be ironed out in the editing. It was a mechanical problem.

Leichtman: Yes. I remember in the book *Raymond* a description of Mr. Myers [F.W.H. Myers, one of the founders of the Society for Psychical Research] coming through a Mrs. Kennedy by automatic writing after his death. He left a message and then said something like: "I won't be back again because you get so nervous when I'm around you." Something like that. Her nervousness prevented clarity of expression.

Sir Oliver: Of course, your rendering of his words was somewhat different than the intent of his words. He was a better-grounded spook than that message seems to indicate. He realized that he was making her very nervous and that it would be better for him to leave and let her find someone she would be more comfortable working with. As a spook, he was more intelligent than that quote makes him sound. He was not being fey or petulant at all.

Leichtman: I suppose that one of the reasons why you often appeared as an anonymous person to mediums was to minimize their nervousness.

Sir Oliver [misunderstanding the question]: Oh, yes. Can you imagine me coming through to give an important message at Madame Zenobia's and announcing myself as Sir Oliver Lodge? Given the consciousness of some of the people who would be surrounding Madame Zenobia, it would be totally ineffective. In psychic circles, my name has the kind of mystique a movie star's name might have. So, if I announced who I was, no one would hear my words. They would be too enchanted by the name itself. So there are many times when I would rather not announce my real name. I have a "dial code" that I use with people that I come through commonly. Of course, when I've come through David in the past I have always used my actual name, because he can handle it.

Leichtman: My question really had to do with going to mediums while you were still alive physically. Didn't you often go anonymously?

Sir Oliver: Oh, yes, but that was because it was the only way to insure that the sitter [the medium] was not going to research me. There are too many mediums who are totally fraudulent; instead of actually contacting spirits, they just make up stories from facts that they have culled from newspaper files about their clients. I'm sure there are a number of films in your time that depict this kind of fakery.

Now, I am not saying that it is at all fraudulent for the medium to know something about either his client or the spirits he's contacting. If a genuine medium knows that in one of his appointments he will be asked to contact Oliver Lodge, for instance, it's perfectly all right to pick up a book on Oliver Lodge and read it. That primes the pump and makes it easier for me to come through. But when a so-called "medium" relies entirely on published data, word of mouth, and information that he has researched—instead of making a legitimate contact with a spirit—then that's fraudulent. That's what I was trying to avoid by appearing anonymously to mediums.

Incidentally, David has read a few short pieces in one or two of my other books, but not very much. I don't think he has even seen a copy of *Raymond.*

Speaking of books, did you know that there is an unknown novel by Arthur? It is still completely in manuscript. As a matter of fact, it was written under a *nom de plume.*

Leichtman: Did Arthur write it or a medium?

Sir Oliver: Arthur wrote it. You would call it a "seance fiction" novel. It is one of the Sherlock Holmes stories, novel length. I don't remember his reasons for withholding it. Perhaps it was a little too sensational in his lights.

Leichtman: Do you suppose that it can be rediscovered yet?

Sir Oliver: Oh, yes: it's going to come to light, but bear in mind that it will be under a *nom de plume*. It is in the possession of one of his descendants and it is known to be a manuscript by him.

Leichtman: Do you still see a lot of Sir Arthur?

Sir Oliver: Oh, yes. We still share many of the same interests. And just for fun, we've been working together a bit on the new wave of Sherlock Holmes stories.

Leichtman: Oh, very good. I think some of the authors of those new stories would be delighted to know that. They do take certain liberties, but—

Sir Oliver: They are great fun, though. Actually, Arthur pays more attention to inspiring those than I do. After all, it's his work in the first place. I guess you could say that I'm just a Watson to his Sherlock Holmes in that department!

[Laughter.]

David asked you to ask me about *The Mystery of Edwin Drood*. [*The Mystery of Edwin Drood* was the last novel by Charles Dickens and was left unfinished by his death. A short while later, an ending to the novel was published by a medium who said that it had been dictated by the spirit of Dickens.] Mr. Dickens did dictate that ending to that person—I believe it was in Canada. It's a valid ending to the novel.

As a matter of fact, many books are written with a certain amount of inspiration from spirits. The author becomes so imbued with what he's writing that spooks can dictate to him quite easily. Yet it doesn't feel any different from the process of writing without such help, unless one is aware of what signs to look for.

Leichtman: I would imagine that it would be very healthy to encourage such assistance from spooks as a part of one's creativity and inspiration.

Sir Oliver: Yes. On the other hand, I don't mean to sound as though all creativity were dictated by spirits. That is not true. When someone is being truly creative, he is tapping into the genius of the cosmos. That's not the same as contacting a spirit guide. A creative person is the kind of medium who can tap the essential creative force of the universe and do something constructive with it. This inspiration can occur directly, without any help from spirits. But sometimes the effort made by the creative person attracts the attention of similarly creative spirits, and they help him a little. But the main part of the work is his own. What I am saying should not be interpreted in such a way as to discredit the work of creative people and geniuses. Indeed, what I am saying is that the creative process is much more than people have given it credit to be.

Leichtman: I understand. May I switch subjects? I would like to know if you are interested in commenting on psychic healing as it is practiced in Spiritualist churches?

Sir Oliver: Hmmm.

Leichtman: Or would you care to pass up that question?

Sir Oliver: No, I'm just chewing my tongue in an effort to come up with some kind of constructive comment.

Leichtman: Many psychics and mediums apparently feel that healing is always a part of any psychic or mediumistic endeavor—that there should always be an attempt to do some physical healing. While healing really isn't classified as a psychic phenomenon, it is often mixed in with psychic activity.

Sir Oliver: I have always wondered why so many people who attempt healing do it with such elaborate fanfare. They seem to be saying, "Look at me—I'm the great healer of the Western world." This type of behavior was common in my day. But to really accomplish a healing, you don't even have to warn your "victim," if I may put it that way. Most of the genuine article is done very quietly, without the patient knowing he's being helped.

There are any number of popular healing methods that are widely used by well meaning but ignorant people—such as "pulling out" or "pushing out" the bad vibrations, "balancing" auras, and so on. In the light of the remarks about psychic viruses already made in this project, and how contagious they can be, it might be a good idea to reconsider what some of these "healers" are actually doing. Many of these people have a "me-

me-me consciousness"; as they set about healing, they may actually remove more health than they give. They sap the sick person by psychically sucking up his loose energy. When a person is ill physically, he is also ill on psychic levels: the illness creates "holes," if I may call them that, in his etheric and astral bodies. As a result, it is quite possible for a "me-me-me" type to psychically hook up to one of these holes and "drink in" whatever energy is available, sometimes to the detriment of the patient. I would like to think that most of these amateur healers *mean* well, but they certainly do not *do* well.

Here again it is valuable to use the measuring stick, "By their acts ye shall know them." These people aren't really healers—they are doing something else altogether. And they should stop and examine what they are doing and why. If they are approaching the act of healing looking for self-aggrandizement or lusting for the everlasting gratitude of others, then they have an improper attitude.

Much of the impetus for the interest in healing among psychics comes from the influence of the Spiritualist Church. And healing in the Spiritualist Church is based on trying to emulate the Christ. Would-be healers might want to remember that Christ didn't go around saying, "I've cured your leprosy, now you owe me a debt." Nor did he say, "Oh, you've got leprosy? Here, let me take a look: mmmmmmmm." *[Sir Oliver then made a slurping noise, as though he were sucking through a giant straw, and smacked his lips. Everyone laughed.]* He didn't cure lepers by sucking up all their unhealthy energy. As a matter of fact, when people do try to heal that way, I imagine that the energy they draw in must taste terrible—figuratively speaking of course. *[Laughter.]*

Leichtman: I have noticed some people who claim to be psychic healers are what I would call "psychic vampires." Just as you described it, they draw out the energy of the sick person. I can't imagine these people being able to heal anything except problems caused by too much energy, such as anxiety or panic.

Sir Oliver: And even then this type of activity is not proper. Stealing someone else's energy is a sure way to attract the attention of the cosmic police. What most people forget is that there is a cosmic police force; when things get too far out of hand, they will take steps to correct them. Within a few years, psychic healers who use vampiristic practices wind up in abject poverty, with the world going on without them. In my own career, I ran into people like this. And since these so-called healers are virtually eating the vibrations of the diseases of the people they are "helping," they usually also become rather sickly themselves. They don't know how to rid themselves of the diseased energy. When one is participating in healing, it is important to learn the knack of not taking on the disease, but this is one of the arts of healing that is not well taught in most circles. If a would-be healer doesn't know this, he can eventually accumulate all of the diseases he has "cured" and suffer all of the pains connected with them. Of course, in most of these cases, these "healers" have abundantly asked for these problems by their attitudes, motives, and activities.

Leichtman: Are you referring just to physical illness now, or also to—

Sir Oliver: To the whole spectrum of illness, which includes mental illness, moral illness, and karmic problems, which are illnesses, too, in a way. This spectrum even includes financial illness. I understand that in the United States there's a very prevalent attitude which considers the lack of finances a sickness something akin to alcoholism—and that there are "healers" who purport to heal individuals of their financial problems. I'm afraid that anyone gullible enough to be attracted to this kind of healer deserves the kind of healing he receives. In some cases, the healing is accomplished by relieving the patients of the excess weight in their pocketbooks. *[Laughter.]* Of course, this is recognized by the due process of law as a criminal act.

Wherever there is a genuine article, be it psychic ability or healing, it will also be possible to find charlatans and mountebanks close by. But always remember that the charlatans and mountebanks are likely to wind up with more in their bags than they had bargained for.

Leichtman: Some healers make a big point of needing a group of other physical people present from whom they can draw energy to use in their healing. Is this a legitimate way to heal?

Sir Oliver: No. Someone who needs to draw energy from a group is not really a healer. He is operating vampiristically, soaking up energy that he cannot obtain in the correct way. A true healer is able to draw from the universal energy pool and then direct it to a diseased person, a diseased part, or a diseased situation. All of these forms of healing are within the province of a good healer. But the vampiristic healer is actually creating illness by drawing away the energies of others—not curing it. He is also creating a karmic problem for himself.

I must tell you this: the amount of energy required to cure even the most dire diseases is so small that I don't understand how this fiction that the healer needs a group of twelve—or whatever it is—to draw energy from ever got started. Even if you didn't know how to tap into the universal energy pool and had to draw on your own energy, true healing takes so little energy that the whole idea of drawing it from a group is absurd. In a sense, the amount of energy used in healing is almost microscopic. I should add, however, that genuine healers find it easier to do their work in the presence of a large group. This is *not* because they draw any energy from the people assembled, but rather because the silent prayers of the group invoke great power from universal sources. Since true healers are agents of this power, these prayers greatly assist them in their efforts.

Incidentally, I think you would find that a psychic who ordinarily charges for a reading but who is also a true healer would not expect or even accept a fee for a healing. In fact, most of the time he would operate without the person being healed knowing that anything was happening.

Leichtman: There is a variety of healer that uses what I call a militantly positive thought—that only health dares exist in this particular body. While this type of healing is apparently effective at times, does it really cure the illness or only drive it down deeper in the subconscious?

Sir Oliver: This isn't true healing; the healer is just using a kind of hypnosis. Hypnosis alone will not heal anything: its purpose is to rechannel the patient's thinking. So, this approach to healing only works if the subject is willing to re-hypnotize his own thinking into a channel of health.

Leichtman: What happens to the pattern of illness after it has been seemingly removed by the hypnosis? Has it been driven deeper—out of the etheric body and into the astral or mental bodies? Does that ever happen?

Sir Oliver: In some cases, yes. Particularly with cancer. Cancer is still a very widely misunderstood disease. In most cases, cancer results from the desire of the inner being to rid itself of the body. Since the cancer is therefore often the choice of the inner being, the militant use of hypnotic suggestion can only repress the symptoms of the disease; it cannot cure it. Even though the disease has been repressed, it continues to fester somewhere in the astral body; eventually, it works itself through to the physical body again.

By comparison, true healing would dissipate the disease on all levels.

Leichtman: Are there cases in which the patterns of illness continue to exist and blossom forth later in a *different* type of illness—perhaps even in a subsequent lifetime?

Sir Oliver: That can happen, but it's a little more rare than you think it is. That kind of repressed disease would tend to reappear in the same incarnation.

Leichtman: Since we're on this subject, what do you think of attempts to heal one's emotional—or say financial—problems by using militant positive thought?

Sir Oliver: Positive thought does have its place in healing. Don't ever suppose that it should be completely done away with. But positive thinking also has to be reinforced by positive activity and the rechanneling of energies. When it's not, it tends to be too superficial to accomplish much healing. Unfortunately, people tend to think rather simplistically, and black is not always black and white is not always white when it comes to healing.

For example, let's suppose that you developed a migraine. How many people that you know would be likely to confront you and say, "You can't be much of a metaphysician because you have a migraine"? Well, such people are not accepting the fact that you're having a migraine which is demanding so much of your attention that it's difficult for you to overcome the pain sufficiently to do the healing. Even an adept who had a migraine would have to go to another adept for healing, because his attention would be so wrapped up in the pain. The positive thinker who is condemning you because you are letting the migraine continue is failing in his duty; he should help you get rid of it first and then perhaps work with you to find out why the migraine occurred. There are sometimes good reasons why migraines occur—even among the most advanced metaphysicians.

Leichtman: Unless you have some more comments on healing, I would like to change the subject.

Sir Oliver: No, that will do for now.

Leichtman: I wish to ask you a question about the puzzling variations in the reports about the nature of the afterlife that come from spirits speaking through mediums. Even when very enlightened people pass over, they often describe their conditions in one way at first and then change the description later on.

Sir Oliver: That's a question my son Raymond would like to answer, so if you will give me a minute, I'll switch places with him.

Leichtman: All right.

[There followed a short pause while Sir Oliver withdrew from David's body and Raymond Lodge entered.]

Raymond Lodge: Your question is well taken; the inconsistencies occur because almost everybody goes through a number of stages after passing over, you see. When you first pass over, you usually enter a state of awareness that is sometimes called the "vestibule." During this period, you usually experience the kind of afterlife that you have always expected, due to your conditioning, your religious upbringing and training, and your mode of thought. As long as you are in the vestibule, these expectations have a kind of reality. In an ultimate sense they are not real, but they are real enough during this period of transition.

The purpose of the time spent in the vestibule is to give the transition from the outer planes to the inner planes more continuity. Even though any given person lives more of his lifetime on the inner planes than on the physical, the transition of death is still very abrupt. It can be almost as frustrating as the transition involved in entering the physical plane; in that case, the frustration is being unable to function as you've been used to, because you are now a tiny baby. But in the transition at the time of death, the vestibule experience permits you to continue your patterns of thought about life and afterlife—until you begin to see that these beliefs are all made out of tissue paper and smoke, as it were.

I want to distinguish between the vestibule and the portion of the astral plane that some of the others participating in the project have referred to as the "Peter Pan department." The vestibule is not necessarily a fantasy, but it's not quite real either. It is a kind of layer in one of the inner planes that helps you lessen the shock of passing over. Again I'll use the analogy of a baby just entering physical life: the baby sleeps a great deal during the first years of its life. During these sleep periods, the spirit of the baby is returning to the inner planes where it can be a total being. So, these long periods of sleep help the baby make its transition and become accustomed gradually to its new life. The vestibule is a period in which a person who has recently died gets used to being just a spirit. It's really not a fantasy, because that sounds as if there are lots of little munchkins running around putting up fake scenery to fool everyone with, and that isn't quite what happens. The vestibule experience is a necessary part of nature. Even house plants and pets experience a continuation of their physical lives for a time when they transit. Cats, for instance, have a tendency to briefly go to a vestibule where there are no dogs, where they can enjoy the things that cats really like.

Eventually, of course, you begin to realize that the vestibule is only a vestibule, made of tissue paper and smoke and cotton candy—it's not real. This realization is, in a sense, what is usually meant by the term "the second death." You give up your cherished connections with the physical life and begin to adjust to life on the inner planes *as it really is*—which is startlingly different. If the transition to full reality were abrupt, and you weren't prepared for it, it would be quite a jolt. Something akin to a psychiatric problem could develop. I believe that it has been mentioned that Colene's grandmother and grandfather [Colene Johnson, David's wife] were given the chance to have something of a second honeymoon in a part of the vestibule before getting down to the real business of living on the other side. I think that this was described as happening in the Peter Pan department, but it was really the vestibule. So, that can be another purpose served by the vestibule. There is also a certain kind of healing that must be done in this time right after death. This healing is much easier when the person is in somewhat familiar surroundings and has a sense of continuity with his physical life.

When I was contacted after my death, I gave a number of descriptions about what I was experiencing at the time. These were descriptions of the vestibule. Obviously, after someone passes out of the vestibule, his descriptions of the afterlife will greatly change.

Leichtman: Are there some people that get caught up in this vestibule for an inordinately long time?

Raymond: I suppose you are thinking of the common phenomenon of a ghost "walking the battlement." Ghosts are people who have died but who are still very much associated with the physical plane for some reason. They usually die under circumstances that create a great somatic shock—like being killed suddenly or murdered. If people die as a result of a very deep shock, when they regain consciousness they often have the sensation of still having their physical fingers and toes, their clothing and hair—of still being physically alive. Although they have technically passed over, they continue on for a time believing that they are still physically alive. And so they continue doing the same things in the same physical locations as before. If physical people clairvoyantly see them while this is going on, they appear to be ghosts.

For example, if you were suddenly hit in the head with a battle ax and died, you would "awake" from the blow on the inner planes sometime later, but you would probably not be aware that you were dead at all—because death was unexpected. So, you would "walk the battlements"—you would continue doing the things you would normally do. The only change you would notice would be that people don't react to what you do or even seem to see you.

This can be a period of great confusion. Sometimes these people have to be left alone, because this confusion is a part of the life plan that they worked out for themselves—believe it or not. And so they "walk the battlements" for a period of time. But the time spent in this way is never really that long in terms of time on the inner planes.

There is, however, another reason for the appearance of ghosts, which I'm going to add here to save confusion. After a person has made the adjustment out of the vestibule, it's still possible for him to appear in a setting that he knew in life and to be more or less visible to the physical inhabitants, depending on how sensitive they are. In fact, one of the most famous of these "ghosts" was investigated by Arthur and my father; she's still in residence in her house in England. She can be seen by people who are sensitive enough. This person is not a ghost that will not let go of earth; she has something that she is doing. This is not an accident. It is rather complicated, and I would not be allowed to go into it, because that would be tantamount to snooping.

Leichtman: While we are on the subject, would you care to comment on the good or harm that bereaved relatives can do for a person who has recently passed over?

Raymond: Bereavement works two ways, you know. I remember after I died and I was being contacted by my family. I was bereaved, too. I was separated from people that I was very close to, and in the beginning the contacts I had with them served to calm me down a little bit, too.

If someone is given over to grief at the passing of a relative, then it is certainly beneficial on both sides to make contact through a medium. But that kind of contact should be made only as necessary in the short period of time after death. As has been said, if someone insists on talking every Thursday to the spirit of Aunt Tillie, they are doing Aunt Tillie a great disservice. Aunt Tillie is alive and well on the inner planes and has her work to do.

In my case, I was given a rather important job to do, and so the connections with my family lasted over a long period of time. This has been true in other cases as well. As a matter of fact, I assisted Betty White when she was coming through her friend Joan [which is recorded by Stewart Edward White in his book, *The Unobstructed Universe.*] I knew the ropes a little better than she did at that time. She called me by a different name, but no matter. I am known by several other names over here—after all, I have had several in various lifetimes. I did not speak through Joan then—it was not necessary.

Betty passed through the vestibule stage very quickly; it was important to her not to remain very long. She was able to make a quick transition because she had been a medium herself and was already familiar with the inner planes consciously. If you will notice the time that elapsed between her passing and the first contact she made through Joan, I believe you'll find that it was a little over a year. [Actually, it was only a half year.]

The time period when I was being contacted after my death covered many months and included several gaps. During this time, as I was describing what was happening to me, I was beginning to make the transition out of the vestibule. And so one account might be quite different from an earlier one, because I had moved on to another stage. Of course, if I were to describe the conditions where I am now it would sound entirely different from anything published in *Raymond*.

The old idea that "in my Father's house there are many mansions" applies to the inner planes, too, you know. You might live in California during one stage of your physical life, and then move on to another location later on. You can do the equivalent of that on the inner planes, not by going some place faraway, but by going on to a different stage. The scenery varies in the sense that the conditions of life change. I don't mean these comments to be taken literally.

Leichtman: I understand. What would you say would be the ideal attitude for relatives to have toward their departed loved ones—the ideal attitude and activity?

Raymond: Once you are convinced that Grandma is alive and well on the inner planes, you should let go of your grief so she can live in peace. Having been reassured that all is well, you should accept the fact that she is now living in a different dimension of life and has new activities to devote her attention to. And so you should try not to interfere with those activities. Of course, if after a few years you need to talk to Grandma, then it is quite possible that Grandma will come back and talk with you. But that should be after Grandma has made her transition and adjustment, and has embarked on her particular work over here.

Leichtman: Is it useful or necessary to send departed friends and relatives love or light?

Raymond: Oh, yes. You know yourself that the spirits that are commonly around you are always sending you love. You bask in it, almost like taking a shower in it. Well, it works the other way around, too. Of course, in your case these are mostly friends in the spirit who are sending you this love, rather than relatives from your current life.

Bear in mind that I am not now the son of my father from

my last lifetime. That was one life episode. Neither am I the child of my parents in the life before that, or the one before that. Those relationships do not carry on, and it is a mistake to expect them to. They carry on for awhile after death: in the case of an elderly married couple who pass on within a few years of each other, for example, it is part of their healing and adjustment to be together again as man and wife briefly in the spirit and have what we can call a second honeymoon. But after they leave the vestibule, they won't think of each other as man and wife in the physical sense. They will, however, probably continue to be aware of each other; after all, their inner beings have probably known each other for a long time. That's why they chose to get married in that specific lifetime.

Family relationships pertain to one given lifetime. I don't mean to disrupt the unity of the physical family with these comments. That's a sacred thing. But it is important to realize that everyone is so much more than the child of his physical parents or the child of his physical parents in his last lifetime or the husband of his wife ten thousand years ago. We can't hold all of these ties, and it was never meant that we should. On the other hand, we do know each other: I frequently talk with people I knew in a lifetime I had in the sixteenth century. We still know each other. We're friends, but we don't think of each other as relatives. I hope I am making this clear.

Leichtman: Oh, yes.

Raymond: After all, physical relationships aren't that relevant to life on the inner planes. But rumor has it that relatives are still somewhat useful in getting born into the physical plane.

Leichtman [laughing]: And ancestors, too.

Raymond: And to learn simple lessons like human love you need parents and siblings and husbands and wives. On the physical plane, these relationships are very important. I do not mean to imply that family unity is nothing: quite the contrary. But when you consider that we live forever, then obviously any one of us has had ten thousand times ten thousand families. That's a lot of relatives!

Leichtman: And a lot of mothers-in-law, too! *[Laughter.]*

Raymond: What I am trying to say is that over here the perspective on physical relationships is different than during physical life.

Leichtman: Are people often helped in making the transition to the inner planes at the time of their physical death? If so, how prevalent is this assistance?

Raymond: Yes, this is done, but it is not a simple task, because those that come to help must be very knowledgeable about what is happening. As I mentioned, for example, there are occasionally beings who make the transition and then "walk the battlements" for several years as part of a meaningful plan. The people who would be involved in helping other beings make the transition must be knowledgeable enough about the whole process of death to allow for the working out of such individualized plans. This is similar to the kind of knowledge a healer must have when he encounters someone with a karmic disease: he must know enough to leave it alone. The karmic patterns cannot be altered. In a sense, "walking the battlements" is a kind of karmic disease.

Leichtman: I understand. I ask this question because there are people who believe that even though they are in physical incarnation, they have a duty to work on both sides of the veil, so to speak. Is this really a common occurrence?

Raymond: I'm sorry—did I understand you correctly? Are you asking about people alive on the physical plane who feel they have a responsibility to help others in their transition?

Leichtman: Yes—while they are out of their bodies at night, when the physical body is asleep.

Raymond: Oh, yes—this can be very important. It's part of becoming aware and developing spiritually—all these good things. At some point, it's very important to learn to consciously manipulate one's awareness on both sides of the veil while in the physical body. However, too many people who think they are doing this make the mistake of playing astral games at night and while they are meditating. Of course, this is very detrimental to growth. In many cases it is the reason why people end up walking the battlement, although it is not the only reason. I'm talking about people who get out of their bodies and travel through the astral for the purpose of astrally manipulating other physical people. I am especially referring to people who practice witchcraft—whose method of getting into the astral is to throw a temper tantrum or something very akin to that. Once they get into the astral, they do a great deal of damage to other physical people—and leave an awful lot of sludge in the lower astral plane.

I am being told to repeat that. At this time, many of the people in the world are becoming very concerned about the ecology of the planet. If half of the people who paddle around the astral at night would take a good look at the garbage that they and others have left on the astral, astral ecology would quickly be recognized as more important than physical ecology. After all, the condition of the astral has a very great effect on the quality of physical life. And when a person goes around psychically attacking and manipulating other people, it's the equivalent of leaving tin cans and garbage all over the astral. It's terrible. And, of course, the garbage just stays there and continues to infect people.

Leichtman: Do the people who so devotedly practice astral travel by and large get into the vestibule?

Raymond: Many of them are getting into the vestibule and are learning about the structure of the astral plane. They learn the mechanics of the astral plane and sometimes go beyond that into one of the other planes. We must credit the people who are learning from this effort. But there are many, many people who think that the astral is the only place to go—that that's where it's all at and that that's all there is on the inner planes. Of course, that's not the case at all. The astral is just one plane of existence. I'm sorry, but I've lost the question.

Leichtman: Where do physical people really go when they travel out of the body?

Raymond: Certain self-centered types—

Leichtman: Do they get into the Peter Pan department?

Raymond: Yes—particularly people who use some kind of a drug to "take off," if I may put it that way. The people who have to descend that low to become psychic, never really understanding what it's all about, get into the Peter Pan department or the astral's subbasement—what I call the witchcraft bargain basement.

What most people don't realize is that the lower astral—where most people go first in their attempts to travel astrally—is full of illusions. Some of it's very glamorous, some of it's quite frightening. But it's all illusion. The lower astral plane consists of the same kind of symbolism that is found in the subconscious mind, because that is where the subconscious has the bulk of its existence. So, most of the people who travel astrally are just swimming around in the group subconscious mind of the human race. Personally, I find that kind of activity worse than taking a bath in somebody else's bath water—if you follow that.

Leichtman: A phrase I often use.

Raymond: I borrowed it from you.

Leichtman: Do you mean that ghosts have "people writers"? *[Much laughter.]*

Raymond: No, no. You have ghost writers—we just borrow from the author. *[More laughter.]*

Leichtman: Very good. You've mentioned that spirits get very busy working on various projects after they leave the vestibule. Would you care to comment on the type of work you are now doing on the inner planes?

Raymond: Of course, I'm the second person you've talked to on this particular portion of the program; I am not doing the same thing as Sir Oliver Lodge. My particular activity is a little bit different. At the moment, I am preparing to reincarnate. I'm not going to give you a date or location, or any of the particulars of my coming birth, because that would be tantamount to running down the street naked. I am hoping to be involved in bringing astrology and astronomy back together again; this is planned for sometime about the end of the century.

I'll have to let the other one back in now, so that he can talk about what he's doing, too.

Leichtman: Very good. Thanks for coming.

[*Raymond Lodge then withdrew from David's body so that Sir Oliver could return and resume the conversation.*]

Sir Oliver: I hope this shifting isn't too confusing for the people who will be reading the interview.

My work at present is many faceted. For one thing, I'm attempting to catch people who are beginning to develop some psychic awareness, but who get attracted to silly practices, such as touching each other and holding hands in healing circles. You know all about these pseudo practices—they are popular in the Spiritualist Church and other groups. I figuratively tweak these people by the ear and lead them onto something better. There are many of us working on this. I also tweak some of the powers-that-be in the Spiritualist Church. I don't mean to be insulting in saying this, but we are in a position to tweak them so that they can grow—which is something that they are going to have to do.

I frequently appear as a control for a medium—it's not David, but it's a medium who does something similar to what we're doing here, although we're not writing a book. It's a trance medium. I don't use the name "Oliver Lodge" because it would not be appropriate in that group. As a matter of fact, I appear as a female by borrowing from one of my incarnations as a woman.

Because of the current interest in psychic matters, my work is focused on trying to help people do something more profound with their psychic gifts. Several of us are working in these ways.

I don't think I'm going to reincarnate any time in the near future, but I plan to make my presence felt in the physical plane in a number of areas. I am certainly not confining my activities to one medium. After all, I'm nobody's slave.

Leichtman: You mean, spooks have rights? [*Laughter.*]

Sir Oliver: Of course. If you think about it for a minute, your "Bill of Rights" is really the printout on the physical plane of the order and conditions of life on the inner planes. I think Thomas Jefferson would agree with me that the physical printout is sometimes incomplete and unsatisfactory, unfortunately, but at least it is an effort to manifest the rights of the inner planes.

Leichtman: Does your work as a spirit also include scientific endeavors? We haven't touched much on your scientific career yet, but I do appreciate the fact that you were one of the preeminent scientists of your day. Are you continuing to pursue these interests?

Sir Oliver: Well, at this point in time, I am working as the inspiration for a brilliant American physicist. He's working on his own, but a few years from now you will be hearing about him. I'm not going to tell you his name, but when you read about him, you'll know who he is. We're working on something that's a kind of De la Warr box. [An experimental physical device designed to project psychic energies for the purpose of healing. It is also known as the Hieronymous device or radionics machine.]

Leichtman: Excellent. Aside from your work in helping this person, are you actively involved in teaching classes, or working with groups of people in the scientific areas of development?

Sir Oliver: In general, yes, but I'm concentrating my main efforts on this one young man. Actually, he has an associate, too—a woman. So, I'm helping both of them.

Leichtman: I have the impression that there is great interest on the other side in helping science discover and apply information about the subtle side of matter and energy.

Sir Oliver: Yes. And in my case, I have the good fortune to be working with a young man who is psychically aware. And the young lady is making strides in becoming psychically aware, too.

Leichtman: Are there many such people—trained scientists who are also somewhat psychic or intuitive?

Sir Oliver: Oh, yes. As a matter of fact, it's very rare to find someone who is developing new inventions or making scientific discoveries who *isn't* somewhat psychic. It goes with the territory.

Leichtman: Would it be possible for such a person, who is able to receive inspiration from the inner planes, to make a major contribution to solving our problems with the energy shortage?

Sir Oliver: Of course. In fact, there's someone in the Southwest who's working on a project that carries through some of Tesla's ideas. He's working on his own. I believe he's developing a power source something like one described in the book *Atlas Shrugged*. In that book, the hero had a car that didn't need any fuel; it had a different power source.

Leichtman: That's very interesting. And it brings up a question that perhaps you can answer. Russia seems to be sending pulsed currents through the earth in some mysterious way, and some people say it is influencing our weather patterns. Are you aware of this?

Sir Oliver: Well, they're not doing it to sabotage your weather. Their scientific community is a bit freer to do certain experiments than the American scientific community is—or the European scientific community, for that matter. They are trying to reproduce some of Tesla's ideas.

Leichtman: Are they being successful?

Sir Oliver: The Russians are getting very close to being successful in these experiments. They are already able to do one aspect of it.

Leichtman: Well, one thing I think they are doing is sending high-frequency electrical waves through the earth. Does this affect the weather? Is it a factor in disturbing our weather patterns?

Sir Oliver: No. Of course, anything that happens on the planet affects the weather, but those transmissions really don't have that much of an effect. I'm told that the reason for the seemingly strange phenomena in your weather right now is just the fact that the world has entered a different climate cycle than it had been in.

Leichtman: Okay. Let me pick up on something you alluded to a few moments ago. Do you think it's more likely for a scientist to make a major creative breakthrough if he is working on his own, as opposed to working within industry or at a university or for the government?

Sir Oliver: This is my own opinion, but I think so. It depends on a number of factors. Some of the scientists in industry are now a little more free to explore creative ideas, and that's good. But to develop some of the really revolutionary concepts of science, it's easier, basically, for a person to do it on his own.

All three of the people I just mentioned are earning a living, are comfortable with that living, and have the time to work on their other projects.

Leichtman: What about the need for immense laboratory facilities? In his day, Nikola Tesla made a great deal of money from a few of his inventions, but always seemed to run out of money to fund his experiments. Is it possible to make many breakthroughs nowadays without huge laboratory facilities?

Sir Oliver: It has always been possible to make certain breakthroughs without huge laboratory facilities. At the beginning, Edison hardly had any facilities. And Tesla did all his initial work without a laboratory.

Leichtman: Well, he did it in his mind.

Sir Oliver: Actually, the problem of tools is really not all that important. There are some discoveries that are made in the fine university scientific laboratories, for instance, but I can think of many more that have come from somebody's garage or rather modest workshop. Often, they come from someone who is not an accredited scientist to begin with—he lacks the education and the degree.

Leichtman: The type of intuitive awareness that such physicists or engineers or chemists develop must be rather significantly different from what the public ordinarily associates with being psychic.

Sir Oliver: That's true. But then, the general public doesn't really know very much at all about the field of being psychic. *[Laughter.]*

Leichtman: Indeed, do even these scientists regard themselves as psychic?

Sir Oliver: The majority of them think of themselves as being intuitive—but it's the same thing. Some of the young scientists coming along now are more aware of their psychic abilities than the scientists of twenty or thirty years ago. They are psychic, are aware of it, and have trained themselves somewhat.

Leichtman: Is there a large body of spirits involved in promoting inspiration through our scientists today? Is this one of the ways God works through our scientists and engineers?

Sir Oliver: It's one of the ways, yes. But it's possible for God to work directly with a person Himself. Most often, however, it's done by someone like me, working closely with a few people at a time.

Leichtman: I asked that question because so many people have the view that divine inspiration comes only through ministers or prophets or priests or mystics. They feel that a person with a highly developed intellect is almost incapable of being inspired.

Sir Oliver: No, psychic ability is a very natural part of human expression. Many people use it every day without realizing what they are doing.

Leichtman: Even intellectuals?

Sir Oliver [chuckling]: Even intellectuals.

Leichtman [sighing with relief]: Well, it's refreshing to hear you say that, because there are so many people who claim that a strong intellect is a *barrier* to intuition. They seem to think that a good intellect is somehow innately antagonistic to the intuitive process.

Sir Oliver: That's an unfortunate attitude. The more fertile and active the intellect is, the more easily a high quality of intuition can come through—if it's allowed to.

Leichtman: What can we look forward to, then, in the next few decades coming from these groups of inspired scientists? Will there be new breakthroughs in the area of energy?

Sir Oliver: Yes, and some of them will be quite amazing. Some of Tesla's ideas will be reproduced in the very near future. And someone will come forth with discoveries that are not based on science, as we ordinarily think of it; they will be based on alchemy. Some principles that were part of the legitimate study of alchemy will come to light. These are very little known in the scientific community or the world at large, but it will be found that they have very practical applications to living.

Leichtman: Will more attention be given to what is called the "etheric" level of matter?

Sir Oliver: Yes, and astral energy, too.

Leichtman: Machines have been developed that can apparently map out and measure the quality and quantity of the etheric or health aura.

Sir Oliver: Are you talking about "L" meters?

Leichtman: Yes. I believe they've been experimenting with these machines at UCLA.

Sir Oliver: The "L" stands for "life."

Leichtman: Yes. They map the "L" fields and "L" lines. Is there any effort being given to the area of scientific acupuncture, to bring through some new discoveries about the true nature of the physical body?

Sir Oliver: Yes, there's quite a bit of research being done right now on the theory behind acupuncture. A tomb in China has just been opened; eventually, this will prove to be a greater find than Tutankhamen's tomb. Among the items found in the tomb is the prototype of an invention that was invented centuries and centuries ago. It's based on a different kind of technology and scientific knowledge than we have today. So it will take about fifteen years for the people who find it to figure out that it is a machine, and what it does.

Dr. Henry Drayton: I've been working with a man in Paris; he's been working with acupuncture for perhaps the last ten years and is particularly adept in this field. I was wondering if there's a team of spirits assisting him from the other side. He has nightly contact with information sources. When he awakes, he writes down the insights he has received and works on them during the day. He's quite psychic, although he will not admit to such phenomena. He's not psychically aware, but he does deal with the information that comes through to him at night.

Sir Oliver: When he publishes some of his material, it will cause a great deal of interest in his theories. He's a harbinger of this period—a harbinger of new discoveries into the nature of acupuncture as a tool and the nature of the etheric body.

Drayton: At the moment, he's dealing with the I Ching trigrams, their mutations and transformations, and using them in his acupuncture medicine for diagnostic purposes. He is therefore dealing with energy in its primordial sense, and how it transforms and changes into an inevitable disease process—and the therapy of that. He hasn't yet gotten involved with etheric energy and its circulation as a means of explaining these processes. Do you see that coming?

Sir Oliver: Oh, yes. I do not understand the I Ching—I have not studied it. But I'm told there's a great deal of wisdom concealed in the I Ching. It is there for those who can see it. The interesting thing is that a number of discoveries are going to be made about the culture that developed acupuncture that will greatly aid the study of acupuncture.

Drayton: Archeological discoveries?

Sir Oliver: Yes, of one sort or another.

Drayton: Out of curiosity, what kind of group on the inner planes is my friend in Paris working with?

Sir Oliver: There's a group of twenty or thirty spirits working with him.

Drayton: He frequently refers to a classical Chinese figure, Chuang-tzu, as being one of his chief mentors among those who come to visit him. [Chuang-tzu was a Chinese philosopher who lived in the fourth century B.C.]

Sir Oliver [laughing]: And this fellow can't bring himself to admit that he's psychic?

Drayton: No, he claims himself to be a scientist and therefore rejects the notion of psychic information.

Sir Oliver: Well, as long as the information is getting through, that's all that really matters. How does he receive the information—is it essentially a kind of automatic writing?

Drayton: No—he'll pose questions before he goes to sleep, or will simply be working on a question in his mind. Then, visions, ideas, and words come to him while he's sleeping, and he'll get up and write them down in the middle of the—

Sir Oliver: That's one of the more common ways people experience psychic inspiration. Many people have received good ideas in that way at one time or another.

[A clock in David's house then struck three. The actual time, however, was 6:15.]

Leichtman: There seems to be absolutely no rationale to that clock. It doesn't even know how to count! *[Laughter.]*

Sir Oliver: I believe Thomas Jefferson suggested that it be exorcised. *[Laughter.]* I can just see David running it out on a leash with its custom-made tennis shoes on—

Leichtman: Oh! Exercising it! *[Laughter.]*

Sir Oliver: Well, even though it can't count, I'm afraid the bell tolls for me. I don't have anything more to say. I would like to conclude my remarks.

Leichtman: I have no further questions.

Sir Oliver: I hope I haven't confused anybody. Has anybody caught an unfinished sentence or thought? Is there anything that needs clarification?

Leichtman: No.

Sir Oliver: Anyone want to throw brickbats? Well, so be it then. Goodnight.

Leichtman: Goodnight.

ALBERT EINSTEIN RETURNS

New ideas arise, in every field of human endeavor, every year. Some are worthwhile, some are not. Most are just variations or opinions of ideas that have already been enunciated. They pile one on top of another, contributing more to the *mass* of human thought and knowledge than to its *quality*. Occasionally, however, a truly great thinker will sweep the idea heap aside with a burst of mental brilliance which not only reforms our understanding of life, but actually revolutionizes the way we think. Such was the impact of the ideas of Albert Einstein, the greatest theoretical scientist of the twentieth century. In 1905, while serving as an examiner in a Swiss patent office, the then-unknown Einstein published not one but *four* papers which revolutionized scientific thinking. The first provided a theoretical explanation of Brownian motion. The second revised science's understanding of light and justified the photoelectric effect. The third and fourth introduced his special theory of relativity and the mathematics supporting it.

Any one of these four papers would have been enough to establish Einstein's reputation scientifically. In fact, it was for his work with light, not his theory of relativity, that Einstein was awarded the Nobel Prize in physics in 1921. The theory of relativity was just too complicated for even most physicists to understand. Even as late as the Twenties, it was frequently remarked that only ten people in the world could understand it—let alone create experiments to validate its accuracy.

Over the past half century, this situation has changed. Physicists have responded to the challenge of these bold new ideas, and now the theory is widely understood by science. Experiments have been conducted which verify the soundness of Einstein's postulates. And relativity has been incorporated into scientific thought, if not yet into ordinary human thinking. Where not so long ago science thought in very static, solid terms, it has now been nudged, if only slightly, toward a more multidimensional comprehension of life. This is perhaps Einstein's greatest contribution of all—not just to science, but to humanity as a whole.

Indeed, Einstein's world was not confined just to atoms, molecules, and mathematics. It was as large as humanity and as broad as civilization. While his fame rests on his achievements with relativity, Einstein was actually quite active in many fields. He lent his name to many humanitarian causes and spoke out courageously on many social issues, from pacifism to the threat of Hitler to the need for more sensible systems of education. Painfully shy, he nevertheless disciplined himself to wear the cloak of a public life, sacrificing his time and energy he might otherwise have devoted to his real love, theoretical physics. Many scientists dismiss these activities of Einstein's as an unproductive diversion from his true work, but this is not the case. Einstein, far better than most of his colleagues, knew that the role of the scientist transcended the work of experimentation and speculation. Like anyone else, a scientist is first and foremost a member of mankind and a contributor to society. He must integrate his science with his responsibilities to human evolution.

Because Albert Einstein demonstrated this ideal so magnificently during his lengthy career, he was a natural choice for inclusion in this series of interviews, *From Heaven to Earth*.

Einstein is truly a priest of God—a priest of scientific investigation. His life is proof that it is possible, even desirable, to be both a scientist and a mystic. In an age when many scientists felt obligated to associate themselves with agnosticism or atheism, as a demonstration of "objectivity," Einstein boldly stated that the universe was an exactly engineered system. "God is subtle but He is not malicious," he said. And he spent his lifetime proving this conviction.

This is the proper occupation of the scientist—to explore the phenomena of life and reveal the patterns and laws by which life functions. Unfortunately, not many of our modern scientists approach their work with this goal in mind. In the effort to be rigorously precise, they have needlessly restricted themselves—and their science—to the study of the obvious, the measurable, and the materialistic. They scoff at the possibility of nonphysical levels of energy. They divorce themselves from the culture in which they live and disclaim responsibility for the consequences of their discoveries. Ironically, by adopting these attitudes toward life, these particular scientists have in essence estranged themselves from the fundamental spirit of openness and inquiry which is the heart of science.

Scientists, as a group, constitute a large portion of the creative, rational mind of humanity. To this vast group is given the responsibility to discover and harness the forces of nature for human use, to explore the universe and comprehend it, and to refine the use and applications of the human mind. The contributions science has made to civilization are enormous. Its understanding of the human physical form has led to great advancements in medicine and the eradication of the devastating plagues of yesteryear. Its development of the basic principles of mechanics has led to the invention of great time-saving machinery and whole new generations of manufacturing equipment. Its discovery of the principles of electricity has led to a transformation, in less than a century, of our lifestyle, our capacity to communicate with one another, and the scope of our activity. Without the breakthroughs of science, the interviews in this series could never have been recorded on tape, transcribed, typeset, printed, or distributed worldwide, as they are. Indeed, the life we are now accustomed to would not exist; life would be devastatingly primitive.

But as significant as these contributions have been, the real value of science is even more subtle yet. As the level of science and technology increases, less of our time individually has to be spent in mere survival. More attention can be given to refining our interests, growing spiritually, and enriching both our personal talents and the culture of our society. The fruits of science and technology make possible a climate in which human consciousness can grow and the mind and the intuition can be stimulated.

Much *has* been accomplished by science through its "holy inquiry," as Einstein called it—and yet so much more waits to be accomplished! The surface of scientific understanding has barely been scratched, let alone penetrated to the core. But many of our modern scientists do not seem to understand this simple and basic point. A peculiar dullness has crept over science today like a veil, stifling the true spirit of discovery. A great many scientists seem comfortably resigned to making more and more refined measurements of less and less, having forgotten that the measurement of phenomena is the work of the *technician*. The role of the *scientist* is to know what data to search for and to make sense of it, once it is collected and measured. While technicians tinker in their laboratories, the real scientist quietly moves into new mental territory and discovers what has always been there, waiting to be recognized by those with the discernment to find it.

Naturally, the true role of the scientist is one of the most significant topics I pursue with Einstein in the interview which follows. Repeatedly, Einstein stresses that scientists need to take a broader view of the world and speculate about the purpose of their work and the consequences of their contributions. He criticizes the intense materialistic focus and self-serving expediency so often found in scientists, and calls for scientists to recognize their duty to lead the way in refining ethics, discovering the inner principles of life, and demonstrating that we do, indeed, live in a rational world, even though popular opinion believes otherwise.

One of the solutions Einstein offers is to enrich the quality of the education and training scientists receive. He points out that the deeper nuclear physicists probe into the phenomena of the nucleus of the atom, the closer they come to touching metaphysical or mystical realities. Some training in the metaphysical aspects of life, he suggests, might help foster a train of thinking which would facilitate scientific discoveries of the inner phenomena of life, in addition to the physical phenomena.

From there, Einstein expands his comments to the education of society as a whole and makes a number of observations about the learning process which produce a great deal of food for thought. He comments as well about the scientific basis for astrology and the nature of time. But Einstein did not just restrict himself to commenting philosophically about weighty issues. He introduced us also to the laboratory of the mind.

Many scientists feel they must work in very expensive laboratories with the most sophisticated equipment. Yet Einstein produced his theory of relativity in the laboratory of his own mind. Now that he lives permanently on the inner dimensions of life, he has found that the laboratory of the mind is an even more sublime place to work than he had imagined before. Free of the dense physical body, he is able to move freely over vast distances and study the full spectrum of matter in all its states, from the very subtle to the gross and back to the subtle again.

To give us some insight into the power of the human mind to study life, Einstein devoted much of his interview to an experiment—an experiment in describing the relationship between matter and life. In a way, it serves as an introductory statement to a new cosmology, a cosmology which is applicable both to the scientist and the esotericist.

His starting point in this experiment is the observation

that modern science knows only of the densest fraction of matter, and that it is oblivious to the bulk of what happens to matter. To support this basic premise, he spends a great deal of time carefully defining the distinctions between life and matter and the various levels of matter. He describes the creation of the solar system, commenting in the process on the validity of the "big bang theory." Throughout, he emphasizes that he is not just speculating on these subjects, but actually reporting on what he can see and observe, at the level of spirit. And he makes it clear that he is not giving us just fascinating but useless facts. He presents us with a key to grasping more about the total universe in which we live and the intangible forces which influence everything around us.

Despite our probing questions, Professor Einstein would give no more than hints about how his comments could be translated by scientists into actual breakthroughs. His purpose, he says, is not to titillate us but to lay out a general statement about the transformation of matter as it affects our lives. At some level, I believe that his description of the movement of new matter into dense forms and then back into more subtle states is a breakthrough in its own right—if it is seized upon by the right people in physics, medicine, and chemistry and used as the basis for further speculation. Nevertheless, Einstein stops short of giving us new formulas for the solving of the problems of modern science. He insists instead that everything that science needs to know in order to make these breakthroughs is perfectly obvious. We just need to learn to observe life more intelligently.

More than anyone I have interviewed in this series, Einstein made it clear that the universe in which he lives now, as a spirit, is anything but a meaningless world of atoms and molecules, interacting randomly. On the contrary, it is a vast, multidimensional realm filled with a full spectrum of matter which faithfully serves the purpose of incarnating life. He is certain that all measurable forms of matter are only the shadow of something greater. Scientists, he claims, must now stop scoffing at the "intangible" and begin appreciating the many intangible forces and substances of life—and their power to influence us. Only then will our knowledge of the universe become complete. Only then will we become masters of the physical phenomena of life.

I will readily admit that the first time I read back through the transcript of this interview, I was not entirely sure it all made sense. This is always a factor when new ideas are presented. Now that I have reread the interview several times, in connection with editing it for publication, I find myself deeply impressed with the quality of thought and insight contained in the comments which follow. I hope the reader will be as equally impressed and inspired.

The medium for the interview was Paul Winters. I am joined in asking questions by my friend and colleague, Carl Japikse.

Einstein: Good afternoon, my friends. It's very nice to be here.

Leichtman: It's good to have you here.

Einstein: Indeed, it's very pleasurable. I haven't done this sort of thing for quite a while.

Leichtman: What would you like to talk about?

Einstein: Well, that's pretty much an open question. It may take me a few minutes to get my bearings.

Leichtman: Sure. *[To Carl]* Do you suppose he's ready for all the test questions we're going to spring on him?

Einstein: To make sure it's really me?

Japikse: Yes. How many hairs *did* you have on your head? *[Laughter.]*

Einstein: Would you like me to specify the dimensions of each hair as well? *[More laughter.]* I would like my contribution to this series to explore the role of the scientist in viewing and participating in the evolution of the planet. From what I've gathered, this seems to be a key precept and basis for each of the interviews. Is that correct?

Leichtman: Yes—participating in God's work in one way or another, as a physician, a horticulturist, a writer, a scientist, or whatever.

Einstein: Well, one of the major roles of the scientist is to view, analyze, and begin to understand the laws governing the manifestation of life on the physical plane—and other planes, too, but for now I will confine my comments to the physical plane. I will try to describe how divine forces influence and govern physical matter and how the ideas of the Creator move into the worlds of manifestation. I think I will begin there, and then we can proceed on to other topics.

Japikse: Other matters, as it were.

Einstein: That too.

Leichtman: Yes, that sounds like a good way to proceed. Of course, somewhere in this interview we want you to give a two-sentence summary of the theory of relativity. *[Laughter.]*

Einstein: Why sure—I'd be happy to. Of course, the tape recorder may have some difficulty picking it up. *[More laughter.]*

Let's begin, then, with the premise that the work of a scientist—and this was certainly true for me—is involved in understanding physical laws and physical manifestation. But understanding these laws requires a willingness to look at them in a fuller perspective than many scientists do. The truly successful scientists are those who take into account something more than what is occurring physically—they also take into account the divine perspective.

Not too many do this, however. In fact, very few scientists take into account the divine perspective of physical manifestation. Most of them focus only on the physical aspects and do not try to understand why there is physical existence—and what physical life means in the greater scheme of things.

Leichtman: Well, why is there physical existence?

Einstein: One of the major and all-encompassing reasons for physical existence is so that life can gain experience in manipulating and dealing in denser matter. It's akin to throwing a young child into a swimming pool so that the child can learn to swim. You watch it so it doesn't drown, but the purpose is for the child to learn to swim. In a sense, God has thrown Himself into physical matter and is attempting to learn to swim in it.

Leichtman: A beautiful analogy.

Einstein: Now, there are many imperfections in physical matter. Many of the divine ideas have not been developed *in physical matter* fully to their final form. They are evolving toward the perfect idea, but they have not reached that stage yet. When the child is first thrown into the swimming pool, after all, it is quite enough that it thrashes about and stays afloat by doing the dog paddle. But eventually, the child would want to learn something better than the dog paddle—the perfect breast stroke, or whatever.

In the same way, the life of God is thrashing about in physical matter. The concept of perfection for physical manifestation does exist in the mind of God, but what is actually being manifested through physical substance is less than perfect. It is moving toward perfection.

The role of the scientist, with this in mind, is to try to understand not just the physical perspective, not just the exact physical properties of substance, but the "cosmic Idea behind the physical manifestation." He should try to penetrate the outer appearance and reach the core of this cosmic Idea, and then become an agent assisting in the transfer of this cosmic design from the inner realms of life to the physical plane. The role of the scientist is to aid in the manifestation of the cosmic Idea. To do this, the scientist must begin by learning as much as possible about the physical manifestation of the Idea, but then he must go on, and seize his responsibility to aid in the evolution of the Idea physically toward perfection.

Japikse: Not just study evolution, but aid it.

Einstein: Yes.

Leichtman: How does the average scientist proceed in investigating the divine patterns of these cosmic Ideas?

Einstein: I'm not finding all the words and concepts I want to use to answer that here in this subconscious [of the medium], but I will do the best I can. The scientist should work with an appreciation of the fact that the manifestation of an idea in the physical realm is the culmination of a trip through many layers of progressively denser matter. Imagine, if you will, a balloon dropping toward the earth from outer space. At the beginning of its descent, the balloon is in space, which is virtually empty of matter as we know it on earth. For the sake of this illustration, let's assume the balloon is floating downward through extremely fine, low density air. As the balloon travels toward the earth, it enters the earth's atmosphere and moves slowly down toward the surface, becoming more dense, more dense, and more dense, landing eventually on the surface. Let's say it lands in the ocean, which would be even more dense than the air, and proceeds to sink down to the bottom of the ocean, which would be the most dense.

Now, if a scientist were to look at the shape of the balloon for the first time after it has reached the bottom of the ocean, he would have a hard time recognizing it as a balloon. It would be collapsed and a great deal smaller than its original size in outer space, because of the effect of the pressure of the air and then the water on its surface. It has journeyed through a whole cycle of experiences from cold to hot, from outer space to atmosphere to water.

The average scientist, I'm afraid, would do a total analysis of the balloon as he found it on the bottom of the ocean and stop there, never once realizing that the balloon had begun somewhere else and had traveled through a whole spectrum of experiences. To investigate the full history of the balloon, he would have to expand his mental faculties and trace the movement of the balloon back along its route of descent—up to shallow water, out of the water into the atmosphere, and back through the layers of the atmosphere to outer space, where it began. Only then would he be able to discover the true nature of the balloon, and all the transformations it has undergone.

Now, to apply this analogy to dense matter, many of the

physical phenomena of this planet are but the physical manifestations of an Idea emanating from God. These Ideas emerge from a central, core Idea which could be, for want of a better term, called pure thought. As the manifesting Ideas descend to earth from their origin, they are clothed in various layers of denser and denser matter, until finally they become apparent in physical manifestation as a form or a law governing form.

To use the human being as an example of what I am describing, the central Idea is the human being's inner self. The physical body and personality are the apparent manifestations of this inner self. Certainly a full understanding of our physical manifestation is important and absolutely necessary, as a beginning step toward self-awareness. But the learning process stops unless we go beyond the physical and assume that there is more to life than our physical appearance. There are energy levels other than physical energy. These need to be explored, too—traced backward along the evolutionary trail.

The scientist can take a big step in learning about these inner dimensions of energy by learning more about himself and the subtleties of his own inner being. What he learns in this way can then be transposed to the larger picture.

It is safe to say, after all, that the physical planet Earth is just the physical body of a certain Being. The physical laws and physical properties of the planet are therefore the characteristics of that Being. Just as a human being has a physical body with a circulatory system and organs and eyes and other features, the physical body of the Being which is our planet has lakes and streams and mountains and gravity and physical laws. These are all characteristics of the physical manifestation of this Being. If a scientist could appreciate this basic premise, it would go a long way toward explaining the origin and evolution of the physical manifestation of the planet.

Leichtman: That's a very good explanation.

Einstein: It may be a little too much.

Leichtman: No, it should stimulate a rich train of thought in those receptive to the kind of inquiry you describe. You are implying, then, that the enlightened scientist would study the physical phenomena of the laboratory and life, but eventually go beyond that and tap the nonphysical components of reality—and even the dynamic, unfolding life behind them.

Einstein: Yes.

Leichtman: Giving an extra dimension to the meaning of scientific inquiry and, I presume, a wholly new perspective on the application of his work.

Einstein: I would hope so. I would hope that by looking within himself, the scientist would see the human family as part of the body of a greater Being. Then, using the laws of analogy, he would begin to explore the nature of this greater Life and the concept that the physical planet is the physical manifestation of this Being.

This kind of enlightened perspective would revolutionize science if it became more common. The results would be astounding.

Leichtman: You are suggesting, then, that today's scientists are very much like a primitive dweller in the Arctic who has never seen water in any form other than snow and big blocks of ice. Such a person would probably have a very difficult time imagining water vapor or steam or a beautiful waterfall, let alone perceive that the clouds in the sky had anything to do with the chunks of ice on the ground. I suppose modern scientists who study matter and physics have just as much difficulty imagining the antecedents of matter and physical laws as the dweller in the Arctic would have problems comprehending the many forms of water. Is that part of the practical significance of discussing matter in this way?

Einstein: In order to understand matter, you must have the wider view of why matter exists, and what its purpose is. Scientists often get involved in studying matter for its own sake, trying to analyze its various properties and deduce from the nature of matter the laws which govern it. And they make no attempt to understand the grand design or purpose of matter. Yet if they would study first the purpose and inner workings of life, the study of matter and the role it plays in life would be much easier.

Leichtman [laughing]: So tell us all about it. Don't just sit there looking smug.

Einstein: I thought you were going to ask questions.

Leichtman: Oh, all right. *[Laughter.]* Why don't we spend some time on the philosophy of the manifestation of life then. We can get to the details later on. You seem to be suggesting that the significance of matter is tied up with the phenomena of life.

Einstein: Yes. Matter is the vehicle through which life is evolving and expressing itself—the cloak, so to speak, that spirit wraps itself in so it can express itself through the various levels of being. Matter provides vehicles for expression and the evolution of life.

Leichtman: Why can't life appear all by itself? Why does it need matter?

Einstein: Life does *exist* irrespective of the existence of matter. But it cannot *manifest* without matter. The descent of spirit into matter is the means by which evolution takes place.

Of course, you probably want to know why this is all taking place.

Leichtman: I don't think we can ask fundamental questions like that in an interview of this nature.

Einstein: That's very true. At the level and perspective at which we exist, I don't think it is possible to know why spirit enters matter.

Japikse: I always thought it was so spirit could have a good time. *[Laughter.]*

Leichtman: Of course.

Einstein: But that begs the question: what is a good time from the perspective of God? *[More laughter.]*

Leichtman: Ah, we're back to unanswerable questions.

Einstein: Yes. All we can do at the level our minds can reach presently is to appreciate that matter is used for the appearance of life. The Eskimo living at the North Pole can only view water in certain ways. In the same manner, we who are living and evolving in matter can only view the issue of matter and the life which inhabits it up to a certain level of understanding. After that point is achieved, the scope of comprehension required transcends the level of comprehension we are capable of achieving.

Leichtman: Well, is it possible to view the phenomena of life separate from the phenomena and movement of matter?

Einstein: Oh, of course. But science is necessarily the study of matter. Science in its pure form is the study of phenomena, not life.

Leichtman: How is it possible to study the phenomena of life itself independent of the matter it is associated with?

Einstein: Well, matter itself would be the mere *representation* and *manifestation* of life. So if we were to try to study life separate from matter, we would have to search inward and explore the subtle planes.

I think we are running into definitional problems here. I have been trying to keep the definition of matter simple, but I may not be able to keep it that way any longer. Let me try to clarify what we are talking about.

If you limit your definition of matter to *physical* matter, as the scientist who studies physical atoms and their relationships and the laws which govern them does, then the study of life *is* far more encompassing than the mere study of the reactions of matter. Life consists of many levels of matter other than physical matter. To use your terms, there is astral matter and mental matter—finer matters which are used in the expression of life *[see glossary for definitions]*. So it is abundantly clear that life can be studied apart from *physical* matter. You simply study the next level of matter—life on the astral plane. Life there is still intertwined with matter, but it is no longer physical matter.

There is a problem, however, in studying life and matter separately if you include all levels of matter in your definition of it—astral and mental matter as well as physical matter. You can separate out the different levels of matter, but not life and matter. Life is enshrouded in matter, and it is not possible to separate them out, except philosophically.

Leichtman: And yet somehow it seems worthwhile to try.

Einstein: Well, yes, but by separating them as different levels of manifested phenomena. You can say, for instance, that there is more to life than physical matter, so you separate physical matter from the rest of life. You can see that life is separate from physical matter; it's involved with it and evolving through physical matter, but can be found apart from dense physical substance. But *where* is it found when not in physical substance? Well, it is found on the next higher level of manifestation, which would be the astral plane. And then you say, but there is more to life than astral matter; this, too, is a manifestation of life. And so you separate astral matter from the rest of life, and move on to the next level of manifestation. In this fashion, you can work your way through all of the subtle planes of existence, but each time you separate the phenomena of matter from the phenomena of life, you run into the problem that the next subtle level of life is involved in the matter of that particular plane.

Are you following what I am saying?

Leichtman: Very clearly.

Einstein: So the distinction between life and matter is dependent on the level of your perspective. It can be separated out intellectually so that you are able to grasp a better understanding of the relationship between the two, but it cannot really be *studied* separately, not in the fullest sense.

Leichtman: Well, let's just take this level by level and see if I have it clear. In the physical dimension, for instance, Carl and I can look at the physical body of the medium through which you are speaking. It is obviously very tangible. We can see it. It moves. It speaks. It is an example of tangible, dense physical matter in motion. But we can also appreciate the equally obvious fact that there is a nonphysical intelligence manipulating this physical form. Would this be one way to separate intellectually the life force from matter, at the level of the physical plane?

Einstein: Sure. Were I and the medium each to remove our presence, the physical matter would still exist, but it would begin to decompose. The physical body here would be lifeless. It would die.

Leichtman: Okay. So a scientist, then, could look for evidence of the intelligence which manipulates the physical matter as a way of separating the life wave or life force from the matter through which it manifests?

Einstein: Yes.

Leichtman: Good. Now, let's suppose that you were a high-powered salesman and were radiating warmth and charm at us while attempting to sell us something. Carl and I would be able to sense this warmth and charm, but we might also begin to wonder about your motive for expressing this affection. Is it natural affection, or just a device for selling us something we don't want? Would this search for the intelligence and motive behind the charm and warm affection be a way of separating the life force from the emotional sensation and matter?

Einstein: Yes, that would be one way of describing it, but you could also continue with the example of the medium. Were the medium and I both to move up to the mental level, permanently, the emotional body would die. The astral matter would continue to exist, but the astral or emotional body would decompose. It would eventually dissipate and return to—

Leichtman: Astral dust?

Einstein: Yes, astral dust. So in one sense now we can look at life as separate from matter, but in a second sense we must see it as inseparable, because it cannot be viewed or understood except in its relationship and interaction with matter, phenomenally speaking.

Leichtman: I understand. Let me belabor this just a little bit more, then, and take it to the mental level. We are hearing certain ideas from you; we can sense the projection of mental radiation from you and even pick up some of the thoughts around you before they are spoken by you. I understand how this happens very clearly. But sometimes, during these interviews, I begin to speculate about what is going on. I wonder why we are pursuing a particular line of thinking or if there is a special purpose in the sequence and selection of the thoughts that are coming into focus. At times, I even wonder if the ideas being presented are really true—or if they are merely a simplistic distortion of the truth, presented in this way just to get a few points across. *[Laughter.]* Now, isn't that kind of speculation another example of separating the life force from the form and matter of thought? I'm not just dealing with the form of the thoughts, but wondering about their larger perspective, relevance, and truth. I'm trying to trace them back to their origin.

Einstein: Yes. And, if you remove the life force, the thinker, from the thinker's collection of thought, the thoughts would begin to dissolve. They would continue to exist temporarily until their vitality was gone, but they would die. It's like pulling the plug out.

Leichtman: Yes. In some cases, it is almost like shutting off the light in a slide projector. Once the light disappears, the image on the screen disappears. It's not quite that instantaneous when you disconnect the thinker from his thought, but in some cases it would be.

Japikse: Sometimes the television set glows for a little while in the dark, after you turn it off.

Leichtman: Yes, the deterioration is slower. My point in pursuing these examples, I guess, is that if a scientist is looking for the evidence of the life force, he should be looking for intelligent purpose, for motive, for the seed thought inherent in phenomena, shouldn't he? He should be looking for purpose as opposed to—

Einstein: Viewing the phenomena.

Leichtman: And sensation.

Einstein: Yes. He should be looking for the driving force behind and the reason for all the phenomena.

Leichtman: So, in the study of the phenomena of matter, the scientist ought to take into account what is going on and why; what intelligence is directing the evolution of this matter?

Indeed, is there such a thing as intelligence within matter? Does an atom have a consciousness?

Einstein: I agree wholeheartedly. These are the kind of fundamental questions I tried to pursue myself.

Leichtman: Brilliantly, too.

Einstein: Thank you. Yes, it is time for more scientists to tackle such questions, and realize that the atom does have a consciousness, and is directed by a life force, a being. This is not a high level of being, but it is there. It is analogous to a cell in your body. The cell has a consciousness, so to speak, but it is obviously a very low level of consciousness. You can talk to me, but I can't talk to your cells. And yet they have their intelligent functions and perform them.

Leichtman: Right. My millions of highly organized, very special cells are doing a good job in terms of manifesting me.

Einstein: Sure. They all have their purposes and functions and know precisely what they are supposed to do. And they do it.

Leichtman: Is there an analogy in this idea that applies also to atoms of dense physical matter—atoms of iron and carbon and phosphorous and oxygen?

Einstein: They all have their purposes and functions. And they all are the manifestation of an indwelling life force which is using the atom as a vehicle to evolve.

Leichtman: Suppose we took the atoms in a leaf of cabbage. What intelligence is directing this matter?

Einstein: The atoms in a leaf of cabbage? The spirit of the cabbage, I suppose.

Leichtman: Well, what if the leaf was plucked from the cabbage head a week ago?

Japikse: Then you have the spirit of slaw. *[Laughter.]*

Einstein: You mostly have the momentum of the consciousness which was evolving through the cabbage.

Leichtman: Well, suppose the leaf has been dead for a year.

Japikse: Good God! Don't eat it! *[Guffawing.]*

Leichtman: Is there still intelligence in those atoms?

Einstein: There is the primitive intelligence of the atoms themselves. In a sense, they are evolving for their own sake now, even though they have lost the connection with the greater whole.

Leichtman: So matter is innately alive?

Einstein: Yes. Matter has an innate life of its own.

Leichtman: If I should eat the cabbage leaf, then some of the chemicals in that leaf would become involved in my body. The significance of the cabbage matter would be enormously enhanced, because it has entered a human life field.

Einstein: Right.

Leichtman: But even all by itself, the cabbage matter has a livingness and primitive intelligence. Matter is never truly dead.

Einstein: No. It moves from vehicle to vehicle.

Leichtman: Well, what state is it in when it is in between vehicles?

Einstein: It always exists in some form or another. It's never really "in between" vehicles.

Leichtman: Are you suggesting that the planet itself can be considered a vehicle?

Einstein: Sure. But don't jump to too many conclusions here. If an atom of cabbage is discarded by the cabbage, or the cabbage dies, it will become part of a different vehicle—maybe a new cabbage or a rock or a human being. But there are restrictions. The best analogy I can use would be to compare the single atom to a tool used in a large shop. Let's say it's a wrench which is used to turn bolts. It would exist in the toolbox as a wrench and would have all the potential of wrenchness, even when it was not being used, but it would only become vital to the work at hand as a worker picked it up and used it. Once the worker was done, the wrench would go back to the tool box and wait for someone else to come and use it.

Matter is something like this. It is used by one particular vehicle over here. When that experience is finished, the matter still exists, but is put back in the tool box, so to speak, to wait for another round of usage. It may be part of a human vehicle at one point and a cabbage at another.

Leichtman: That's all very clear.

Einstein: But there is some specialization. There isn't as much crossing or mixing of different kinds of matter as you might think. There is specialty matter which exists primarily in human vehicles. It "volunteers" to help the human family manifest. And as humans pass from the earth scene, this matter will then be sucked up into the Eternal Void. *[Laughter.]*

Japikse: The "EV," eh? *[More laughter.]*

Leichtman: Yes, we know all about the EV. Well, who keeps this all straight? Who's running the show? It sounds awfully complex.

Japikse: It's all programmed into an Apple computer. *[Laughter.]*

Leichtman: A *big* Apple computer.

Einstein: I guess you could say God—if you wanted a simplistic answer to a simplistic question. *[Laughter.]*

Japikse: Let's assume it wasn't a simplistic question.

Einstein: Ah, you mean it was just a *dumb* question! *[More laughter.]*

Well, perhaps if I talked a bit more about the solar system and the universe as a whole, I will actually be able to give you the kind of answer you are looking for there. I'm not sure I will be able to communicate my thinking on these subjects fully to you; it will be very difficult, patching it through a medium, as it were. But let's give it a try.

Leichtman: I understand the difficulty. We will try to help with the right questions, if the comments don't come through clearly.

Einstein: Thank you. Let me begin by restating my original premise that the transformation of matter occurs as the result of being used by an indwelling life. All phenomena of matter must be seen in this context, if they are to make sense. Now, the best way to explain what I mean by the term "indwelling life" is to say that it is an entity or being which is ensouling itself in matter and evolving through association with matter. It makes no difference whether this entity is a human being, a planet, a solar system, or a galaxy; the principle is the same in each case. Of course, it is very difficult for a human being to comprehend the nature of a being who is ensouling a galaxy, but we can simply state that such a being would be comprised of all the individual units of the galaxy—the planets and solar systems and so on.

Leichtman: I understand.

Einstein: On the human level, the inner being uses physical, emotional, and mental matter to express its inner life. On the planetary level, the inner being of the planet uses physical, emotional, and mental matter for its evolution and manifestation. This planetary inner being exists as a vortex in subtle matter. You can think of this vortex as the intersection of the indwelling life and subtle matter existing in its primal state. As primal matter intersects with and accepts the indwelling life, or cosmic Idea, the indwelling life moves into manifestation as a Being, through the vortex. This is how It becomes a planet.

We can, therefore, discuss creation as the process of this Idea moving through various subtle planes of matter, intersecting them, and accumulating substance from each of the levels, pushing ultimately through to what we call physical existence. On a planetary level, this way of approaching creation would, in effect, rule out the "big bang theory" of creation and sustain the theory in which individual units are formed out of a central molten mass, subsequently separating into various bodies which collectively become the solar system, with its central sun and several planets.

Now let me explain this in more detail. The Idea, which could also be called the cosmic plan of God, sounds forth the creative impulse and proceeds into matter, intercepting with it at various levels, accumulating it at a central core or vortex, and appearing in its initial form as a swirling mass of matter. As the swirling mass of matter expands and grows and solidifies, various secondary vortexes begin to appear and start to attract matter in a similar way. These secondary vortexes eventually become the individual planets.

I am describing the birth of a solar system—at least, it is one way to describe it. The center or heart of the solar system, of course, is the sun. The planetary vortexes can be thought of as children of the sun or separated parts of the sun. Each of them would contain an indwelling and evolving entity which is part of the greater life of the sun. Is this making sense?

Leichtman: Sure. Can we ask some questions?

Einstein: Of course.

Leichtman: What comes first, the matter or the Idea?

Einstein: The Idea comes first.

Leichtman: But the matter must be there for the Idea to manifest, right? Are you suggesting that the matter is preexisting but unorganized?

Einstein: Yes.

Leichtman: Where did it come from?

THE VORTEX

Einstein: The matter came from a separate Idea.

Leichtman: Some people have said that the various grades of matter in the solar system, from subtle to gross, are really the corpse of the previous solar system—its dead body, in effect. This matter has already been partly vivified, but with each succeeding incarnation of the solar system, there is additional vivification and transformation. Is that theory anywhere near the truth?

Einstein: No. Matter existed previous to the birth of the solar system, and it will exist after it is used by the indwelling beings. That is true at every level of being. But the question was: where does this physical matter come from? It was created long ago through a separate act of creation, and left.

Leichtman: It's just used over and over again, without change?

Einstein: No, matter does change, even though it does not evolve. I'm making a distinction here between *change* and *evolution.* It is the indwelling essence in matter which evolves. As this evolution occurs, matter itself is changed. Its quality is transmuted. It is assimilated, transmuted, and left.

Leichtman: Would this be something like the principle of regeneration which can be seen in the cycle of planting wheat, nurturing its growth, harvesting its grain, making it into flour, baking the flour into bread, and then eating the bread? Is it this sort of transformation?

Einstein: Yes, it is very similar to that in principle. A better analogy would be the use of farm land which is continually replanted to grow produce. You would plant one crop for a season, then a different crop the next season, and so on, year after year. It is the same land from year to year, but the quality of the soil changes somewhat from year to year, as different crops are planted. If the crop rotation is managed properly, the soil is enriched. If not, it is depleted. The soil changes in quality, depending on its use. It has been borrowed, so to speak, for the growth or evolution of the crops, but it changes to a certain degree, even though its essential nature remains the same.

Leichtman: I presume that over a long period of time, matter is changed for the better by this association with evolving life.

Einstein: Yes. It becomes more refined in quality, which allows it to be utilized by more evolved beings.

Leichtman: So it mysteriously becomes more responsive to life.

Einstein: Not more responsive, but responsive to a different note, a higher note.

Leichtman: Oh, I see what you mean. It becomes responsive to a higher level of consciousness and can therefore be used to produce a more refined self-expression.

Now, according to esoteric literature, matter in the past millions of years has become more responsive to the principle we recognize as magnetism or cohesiveness. It has become more organizable and more responsive to what we call natural or physical law than it once was.

Einstein: This is true. The entities or beings who are ensouling matter become more organized themselves and more able to organize matter and transmute it into something greater.

Leichtman: These same esoteric sources suggest that the matter of the astral or emotional plane is more highly evolved and responsive than either mental or physical matter, for peculiar reasons.

Einstein: Well, I would not say that. I would say that physical matter would be more evolved because the evolutionary focus in our system is on the inhaled breath rather than the exhaled breath.

Leichtman: What do you mean by inhaled and exhaled "breath"?

Einstein: I would repeat that matter changes as the inner entity or being travels through progressively denser levels of matter and then returns. To go back to our analogy of the balloon, the balloon moves from the rarest or least dense matter into the densest, which would be the bottom of the ocean. But then, to continue the analogy, it would begin to transmute the matter. It would not just stay at the bottom, but begin rising up again, back toward the subtler levels of existence.

The use of the term "breath" is poetic. It is a reference to the breath of God—the act of breathing life into matter. The exhalation of this breath is the cosmic Idea being sent forth into matter. It is the ensouling life force moving into matter. As this wave of life touches the densest levels of matter, it begins a path of return—the so-called "inhaled breath." The ensouling entity is now moving back toward its source and through progressively less dense matter.

Leichtman: I see.

Einstein: So we might say that an entity such as a planetary being begins its evolutionary process by moving from mental matter into astral matter and then into etheric matter and finally into physical matter. In this way, it becomes grounded in physical matter and begins to transmute and vivify the dense matter of that level. As progress is made, it then begins to work its way back through the subtler planes of matter toward its original level of being, having been enriched by the experience of transmuting matter.

Leichtman: You make it sound as if there were two separate streams of evolution. One is the life force of the ensouling entity, and the other is the evolution of matter itself.

Einstein: Yes, except that matter itself is not evolving. It is merely being used. When matter becomes part of the body of a living entity, then it may change somewhat and become enriched, but this change is secondary to the evolution of the ensouling consciousness.

Leichtman: So then, genuine evolution is always in the consciousness of the indwelling life.

Einstein: Rather than the actual matter used, yes.

Leichtman: This seems to imply that the matter we will use to make bodies five million years from now will be more enriched, and that we will be able to do a better job of being humans because our bodies of thoughts, feelings, and physical activity will somehow be more responsive to our soul. Is this correct?

Einstein: Yes. Even on a short-term basis, we are doing

a better job now as opposed to even a couple thousand years ago.

Leichtman: Well, let me ask some questions about the transformation of matter. If I am hearing you correctly, matter is changing in various ways. One change is that the subtle varieties of matter are moving into denser levels, as if there were a continuous precipitation of matter.

Einstein: Yes, there *is* continuous movement toward the denser levels. And as matter is pushed into denser levels, the denser matter of the physical plane is then transmuted into less dense states. There is continuous movement of this nature.

Leichtman: Is this a cyclic movement? Is the matter just changing back into what it once was? Is it just turning around and coming back, from the subtle to the gross and back to the subtle again?

Einstein: No, it is not being recycled.

Leichtman: You mean there is a continuous creation of matter?

Einstein: Absolutely. It is something like a matter machine which is continuously in operation.

Leichtman: Central headquarters is making matter again. *[Laughter.]* Okay, let me retrace what you are saying. If mental matter becomes emotional matter, and emotional matter becomes etheric, and etheric becomes dense physical matter, what is the next phase? Is it the phenomena of radioactivity?

Einstein: No.

Leichtman: Well, then what is the next phase?

Einstein: The dense physical matter becomes a different type of etheric matter.

Leichtman: Well, this is where I am getting lost. It sounds as if it is simply reversing itself, and yet you said that's not the case.

Einstein: In the literal sense of *direction,* it is reversing itself. Having come from the subtle planes, it is now returning to the subtle planes. However, as this dense physical matter is being transformed into etheric matter once again, it goes through a refining process. It is able to be used in much different ways than the last time it was etheric matter.

It becomes *mature* etheric matter rather than *young* etheric matter.

Leichtman: Ah, I see. It is said that in very ancient times certain subplanes of etheric matter were used only by plant life, while humans would use a more refined element of etheric matter in making their bodies. In recent times, things apparently have become more mixed. Is this what you are referring to?

Einstein: In a limited sense. I'm referring to the broad spectrum. Keep in mind the distinction between the ensouling entity and the matter it ensouls. In order to provide a broad cross section of matter suitable for ensoulment, matter itself goes through many changes. There is more refined matter and less refined matter on each of the subplanes.

I have a picture in my mind of a V-shaped diagram. Matter moves down one side of the V and hits the bottom—which is the physical plane—and then it moves up the other side. And on the downward slope it is young matter, and on the upward slope it is old matter. All matter goes through this refining process, moving down through the subtle planes to the point of the V and then back up. Now, all along this spectrum are different types of matter; these differences depend on age and the various qualities it has vibrated in harmony with over its period of existence.

An ensouling entity would use various particles of this matter for the purpose of manifesting its self-expression. Depending upon its evolutionary state, it would pick matter from one level or another. It doesn't necessarily use just one type of matter, however. It may pick some old matter and some new matter and some in-between matter, and put them all together for its own particular purposes.

Leichtman: Would a frog select younger matter for its bodies than, let's say a saint or an angel? I assume a saint or an angel would take the older, more refined, more subtle matter.

Einstein: Of course. You earn the right to use the older matter.

Leichtman: Then the matter which would go into the

physical, etheric, and astral bodies of a frog would be much less responsive to the life force than the matter in the bodies of a saint.

Einstein: That's right. Some matter is moved around very freely to accommodate the various needs of the different grades of ensouling life, but some of it is reserved for special life forms.

Leichtman: Something like matter banks?

Einstein: Yes—you call ahead and reserve matter. *[Laughter.]*

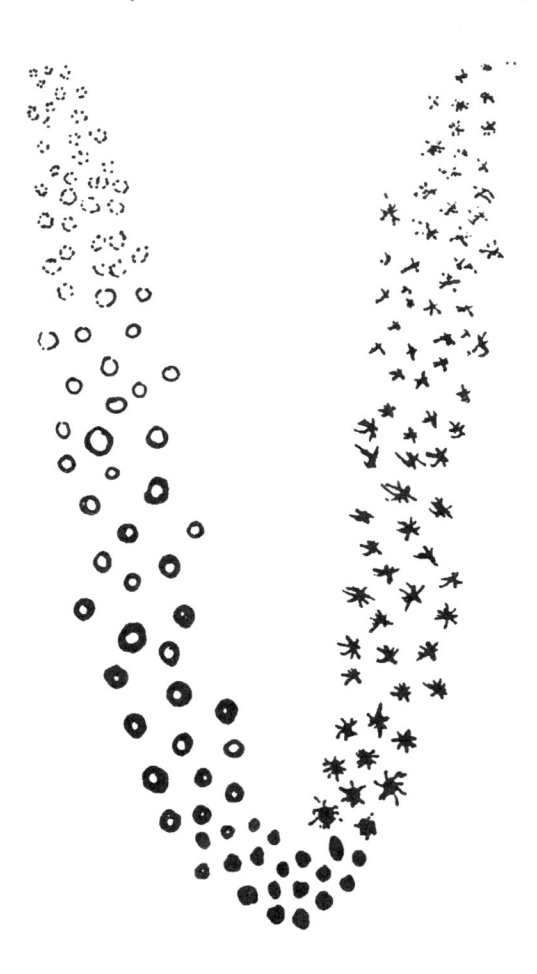

THE ODYSSEY OF MATTER

Leichtman: Like a restaurant reservation. *[More laughter.]*

Einstein: Some of it is removed to other areas. Sometimes particular areas of the universe are producing more refined matter than others, and it is transferred to other places, where it is needed.

Leichtman: So at any one point in time on the physical plane, you would have matter in different stages of development. You would have old and young physical matter. Can you give us an example of what you refer to as younger, undeveloped dense physical matter? For instance, is the carbon in a diamond more evolved than the carbon of wood ashes, both in terms of age and refinement?

Einstein: The best way to detect this refinement is in the way or form the matter is used. Yes, jewels and precious minerals would be examples of the higher quality, more developed matter.

Leichtman: Is the matter which appears in crystallized forms more likely to be more refined than the matter in amorphous forms?

Einstein: Yes. And similar observations can be made about the matter being used by human beings, the more evolved animals, and the more evolved plants. The more evolved the being is who is using the matter, the more refined the matter is likely to be. We cannot measure the actual age or refinement of matter, but we can measure the quality or expression it is associated with. We can measure the functions and uses of the bodies the matter is a part of.

Leichtman: What about the contention made by some people that artificially-made jewels do not have the same properties as the ones made by nature?

Einstein: That's not necessarily a result of the use of old or new matter. And do keep in mind that it is very common to use both in producing bodies.

Also keep in mind that the difference between old and new *physical* matter is not as great as the difference between old and new *mental* matter. The time interval is quite different. If we were to illustrate this spectrum of matter in the shape of a five-inch V, for example, we could then draw a line one inch up from the bottom of the V and say that all matter below this line is physical matter. Now, if the transformation of matter is represented by moving down one side of the V and up the other, and the aging process parallels that movement, then if you were to take new matter from one half inch up one side of the V and old matter from one half inch up the other side, it would be hard to detect too much difference between the two samples.

Yet if we complete this diagram and place astral matter between one and three inches from the base and mental matter between three and five inches from the base, we can readily visualize that the difference between old and new mental matter

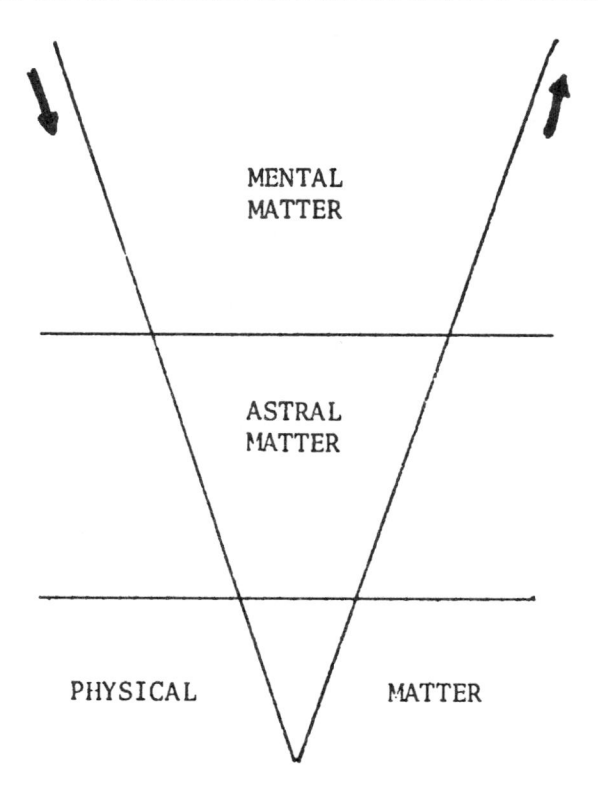

MENTAL MATTER

ASTRAL MATTER

PHYSICAL   MATTER

THE TRANSFORMATION OF MATTER

is far greater than the difference between old and new physical matter. Old mental matter would have traveled through all the various subplanes down to the bottom of the V and then back up the other side.

I suppose the next question is, "So what?"

Leichtman: Well, I *am* wondering about the practical significance of this transformation of matter. If it is connected with our individual evolution and the evolution of the life of the planet and the solar system, as you say it is, then I presume there are facts we need to understand about matter itself. Is there something we can do consciously with matter to accelerate evolution?

Einstein: It will be easier to answer that question in terms of emotional and mental matter, rather than physical matter. A knowledge of the relationship between the different types of matter and the different qualities of emotion and thought could aid in improving the quality of our emotional and mental bodies.

Leichtman: So as you improve the quality of consciousness, you create a corresponding change—

Einstein: Of course.

Leichtman: So again, it is the indwelling consciousness which must initiate the fundamental work of improving the quality of life—meaning us.

Einstein: Yes. The emphasis should be to improve our consciousness, rather than get so involved in matter. At least, this is true in a strictly theoretical sense.

You know, there are many implications to these ideas which will be helpful not just to scientists but to others as well. There are many discoveries which psychologists and physicians need to make, for example, concerning the subtle matter associated with emotions and thoughts. Some of these discoveries will make it easier to discern the difference between the phenomena of consciousness and the phenomena of subtle mental or emotional matter—the difference between the thinker and his thoughts or emotions, to put it in philosophical terms.

Leichtman: There are professional people who do this now—not many of them, but some do talk in these terms. I doubt if very many of them bother pursuing the esoteric details about the subtle matter of emotions or thoughts, however.

Einstein: And it might be a distraction for many of them to do so. I recognize that. But it is also useful to see where this kind of investigation can lead us. One of the discoveries that researchers in psychology and medicine will eventually make some day is that nonferrous matter also has the magnetic properties which ferrous matter does. This includes the matter which goes into the substance of human thought and feeling. It is not the type of magnetism which attracts iron filings, of course, but it is most definitely a species of magnetism. It attracts other substances in harmony with it, as well as repelling matter which is not in harmony with it.

Indeed, there is a whole science of magnetism waiting to be discovered and applied to physical and psychological health. Of course, it will be awhile before it is discovered—science has a lot of homework to do first in comprehending the phenomena of magnetism in its fullness. As long as scientists insist on staying focused in the densest levels of physical matter—dead matter, really—they will miss what I am talking about. The magnetic properties I am referring to here are more likely to be found in the bodies of the higher plants, animals, and humans.

Leichtman: Where should an interested scientist begin in trying to study these possibilities?

Einstein: There are hints here and there in the writings of some occultists and others. Of course, orthodox scientists tend to reject such writings without even considering them.

Japikse: Yes, there will probably even be a few who will reject the comments you are making here, because we are using a *medium!*

Einstein: Imagine that. Well, those scientists usually aren't the ones who make breakthrough discoveries anyway, are they? But that would be one place to look. They might also look up the subject of radiesthesia and consider some of the cures achieved by herbal and homeopathic methods. Medicine is really at the stage where it needs to take a closer look at some of the remarkable results which have come from these unorthodox methods. Certain types of subtle physical or etheric matter seem to attract specific illnesses to the physical body. The right type of magnetism administered as an herbal or homeopathic treatment would be able to dispel or disperse the "bad" matter, leading to a cure.

A parallel situation can be found in psychology. Experimenters will eventually find that emotions must be dealt with both as highly magnetic nonphysical matter and as an aspect of consciousness. The difficulty in treating many emotional illnesses stems, in part, from the fact that the emotions which cause these problems tend to be magnetically responsive to a kind of astral matter which easily "glues" itself both to our own feelings and to more of its own kind. This magnetic action makes it very difficult to get rid of the "bad" astral matter—and the emotional problem.

Leichtman: So where do enlightened physicians and psychologists go from here? How can they investigate these possibilities?

Einstein: This is really out of my field. I can only drop a few suggestions and hints for others to follow up.

A surprising amount has been written on these subjects. These writings are not always easy to find, but the right people are somehow able to find them anyway. Reading these books would help them fill in the gaps in their own thinking and observations. For example, they ought to read about the hidden side of life—the so-called invisible realms and the subtle planes and grades of matter. There is a wealth of material about these subjects which can be investigated scientifically. I'm not talking about probing well-kept esoteric secrets here—just finding out simple facts which have been ignored by science, such as the fact that small amounts of vegetable or mineral matter—the essences of flowers or homeopathic remedies—can have a powerful effect in treating human illness.

Frankly, it is amazing that any physician or psychologist can fail to appreciate that there is an ensouling consciousness which uses the matter of thought, emotion, and dense substance to create the various bodies of manifestation. The very performance of their duties should lead them to try to discern the Idea and inner purpose which is ensouling the forms they treat, so that they can better help this Idea manifest. To put this in pragmatic terms, the physical or emotional body of a person might be sick, but the inner being and inner Idea is quite healthy and seeking to heal the substance, pattern, and function of these sick bodies. That should be a basic premise of medical science. It is a *fundamental* law of life and its manifestation.

Leichtman: It is a rather common concept in many schools of thought.

Einstein: Yes, but it often just remains a theory without much practical significance. There are few medical or psychological techniques which actually incorporate or implement it—or even test it.

I suspect that some of these ideas about the magnetic aspects of physical and emotional sickness will be the focus for some real breakthroughs in medicine, and perhaps inspire similar breakthroughs in other scientific disciplines. In many ways, it is medicine and psychology which will have to lead the rest of the scientific community to a discovery of the ensouling consciousness and its relationship to matter, because they deal more directly with the subtle levels of matter. Of course, the whole field is ripe for many more breakthroughs than have occurred, and some of them potentially could be dangerous. I imagine they will come as the result of trial and error, without a full grasp of why it all works—much as electricity was discovered without knowing much about why it works.

Leichtman: Well, this is fascinating; you've given us some very tantalizing ideas for relating your theoretical comments about the relationship between matter and consciousness to practical applications. I can accept these ideas for several good reasons—not just because they are coming from you, but also because they are consistent with some of my own observations, as well as the observations of others who have written and spoken on these topics. But I would like to go back once more to our discussion of the transformation of matter, from subtle to gross and back to subtle. I am wondering if there isn't some kind of meticulous study of physical matter which will give at least a hint of proof about what you've been describing. Aren't there any sophisticated gadgets or devices which can detect some of the actual changes in dense matter which occur during this transformation?

Einstein: Oh, no. Definitely not.

Leichtman: The changes are too subtle to observe?

Einstein: Too subtle to be observed from a physical perspective. In other words, a truly profound understanding of what is happening in physical matter can only be achieved by viewing it from the next higher level of matter.

Leichtman: So phenomena such as radioactive decay are not part of the evidence which would support this transformation? This is not part of the "maturing" of matter?

Einstein: That's a different phenomenon.

Leichtman: Is there something in the rapidly evolving study of subnuclear particles which would reflect the changes we've been talking about? Would a new look at these tiny particles lead to discoveries which would support your idea of a life force which is sweeping matter from a subtle plane to a denser plane and then back to a more subtle plane? I'm talking about the mesons and that type of particle.

Einstein: All the little guys. *[Laughter.]*

Leichtman: Yes. Would it help the scientists who are studying these phenomena to consider what we've been discussing?

Einstein: In general, I believe not. Let me say this. Being able to prove the transformation of matter through physical experimentation would not really help science achieve what I think should be its basic goal, which is to begin viewing physical phenomena and physical matter as part of a much larger phenomenon.

It would essentially just keep scientists trapped in the physical, validating and verifying their physical observations. It is necessary to run experiments which validate one's observations, but it is extremely limiting to view physical matter only from the physical plane. To understand matter, the physicist must study it from the level of subtle, nonphysical energies.

Leichtman: Well, other than draw analogies from medicine and psychology, how does the physicist do that?

Einstein: He sets up an apparatus to view etheric matter.

Leichtman: Of course. *[Laughter.]*

Japikse: Deus ex machina. *[More laughter.]*

Leichtman: Can you describe this apparatus?

Einstein: Well, he would take a wire from over here and a wire from up there *[motioning with his hands]* and combine them with a power supply. And poof! Out would come the indwelling life. *[Laughter.]*

Leichtman: Is it practical, all kidding aside, for modern electronic engineers or physicists to build devices to study etheric matter?

Einstein: We're getting there.

Leichtman: It's achievable?

Einstein: Of course. It's all a part of the evolution of scientific thinking. It's just that up until now, scientists have basically been limited to studying physical matter. Now they must study the underlying currents and energies which influence physical matter.

Leichtman: Right. Well, this leads to my next question. It seems that the study of matter and subatomic particles has reached a practical limit. In many ways, scientists have gone to the very edge of the physical plane. They are now at a kind of interface between the etheric and the dense physical planes, and they don't quite know what they are looking at. Is there something of the phenomena of the interface itself which could be studied now, leading them on to the further implications we've been discussing?

An analogy would be the study of the interface between

water vapor and liquid water. Obviously, there are certain phenomena which occur at the interface, as the water moves back and forth between its liquid and its gaseous forms.

Einstein: And so, to extend the analogy, you are asking if this can be done at the interface between etheric and dense physical matter. Yes, it can, and I'm sure it is being done. But we are in need of a breakthrough.

Japikse: Is it being done with an awareness that an interface is being studied, or is it being done in the belief that it is part of the known physical level of matter?

Einstein: Oh, I don't think the people working on this suspect that they are at an interface. It's more a case of discovering certain levels of matter which clearly exist but are not observable in any way from the physical level of existence. And the implications of what this means are beginning to creep into the thinking of the people involved.

Leichtman: What line of direction would you suggest for their further exploration? Looking at the phenomena of bevatrons, cyclotrons, and Wilson cloud chambers? Or should they study electricity or magnetic phenomena?

Einstein: Well, I'm not in a position to give you the breakthrough itself, here in this interview.

Japikse: It would sell more books. *[Laughter.]*

Einstein: At least to the handful of people who are actually working on such projects. It wouldn't make much sense to anyone else.

You could go in almost any of the directions you just mentioned and find the interface. The key is studying the qualities and characteristics *behind* the physical movement of atoms. This would reveal that there are subtler particles interacting with the atoms of physical matter.

Leichtman: Are you saying they should speculate on the electromagnetic phenomena within the nucleus of the atom—that some of this is really extremely subtle matter rather than force?

Einstein: Yes.

Leichtman: And that this subtle force somehow turns into matter and that matter turns back into this force?

Einstein: Folds back on itself, so to speak. I couldn't have said it better myself. But keep in mind that it's very easy for us to sit here and say you ought to study this instead of that. In practical terms, it's more a matter of trial and error until we achieve the breakthrough. The work is being done and it is in competent hands.

It is somewhat simplistic to say, yes, you ought to take the nucleus of an atom and examine its subtle matter. Saying that still leaves the practical question of how you do it, of how you develop devices to measure it, and so on. That is what is taking time.

There are many of us here that are pushing the investigators in the right direction. They have not yet made the major breakthrough, but it is coming. It will be a confirmation of the existence of realms of life which are not measurable in physical terms, beginning with the etheric plane.

Leichtman: Okay. I want to ask you about a few physicists who are busy exploring the relationship between certain Eastern philosophies and nuclear physics. There is a man I know, Fritjof Capra, who wrote a relatively popular book called *The Tao of Physics,* in which he speculates on some of these possibilities. The same can be said about another book, *The Dancing Wu Li Masters,* by Gary Zukav.

Einstein: Excellent books. They should be read by everyone with an interest in science. There's a need for more of this kind of speculation. The melding of East and West is, of course, proceeding on schedule. Physicists need to discover the link between the mind and what the mind perceives in physical matter. It has always been assumed, up through the last several years, that the scientist is limited by the restrictions of the physical plane itself. What we must realize is that it is not the physical limitations which restrict science *but the mental limitations of the scientists!* Until this is understood, the major breakthroughs will not be forthcoming.

Leichtman: I would presume, then, that the laboratory the scientist will use to explore these interfaces will not be some monumentally expensive laboratory costing tens of millions of dollars, but simply the imaginative minds of brilliant physicists and their grasp of esoteric mathematics. The language of mathematics, after all, lets you speculate about nonphysical states, fourth-dimensional forms of movement, and all that exciting stuff.

Einstein: That's right. It's particularly important for the scientist now to examine the limitations of his thinking about the physical realm, rather than try to devise more and more complex physical devices.

Leichtman: They can put that cyclotron away and go back to the drawing board. *[Laughter.]* Would it be helpful to study some of the works of the great mystery school founded by Pythagoras?

Einstein: Ah, I see the point you are bringing up, and I will expand on it a little. In ancient times, there was a much greater understanding of the relationship between the mind and physical matter.

Leichtman: You're talking about the ancient antecedents of our present civilization?

Einstein: Yes.

Leichtman: The ancient Greeks and Egyptians?

Einstein: Yes. Although they were not necessarily as sophisticated as we are in dealing with physical matter and machines, they were much more enlightened and efficient in what they did use. Their science was quite sophisticated. An example which is frequently cited is the astonishing achievement of the pyramids. I would love to have some construction company bid on building a pyramid today and see if it would be economically feasible. *[Laughter.]* I'm sure it would not be.

My point is that humanity is not really better off today in our intellectual understanding of matter, in spite of our technology and scientific advances, than we were thousands of years ago.

Leichtman: In fact, many people think our technical advances have made us worse off. Some say, for instance, that all of our excessive mining and drilling for oil is akin to sticking needles and knives into the skin of a dear friend and wounding him.

Einstein: Well, in a sense that is true, but you have to keep the scale of things in mind. On a rational scale, it's something like getting a pin prick in your pinky. It would be annoying but not really harmful.

Leichtman: They use this analogy especially in reference to mining radioactive materials, some of which can become ferocious health hazards in time. Mother Nature dispersed and buried these materials very deep to keep them out of the way of humans, but now we seem to be working contrary to common sense and natural order, by digging them up, refining them, and distributing them all over the place.

Einstein: I don't think these critics should take their analogy so literally. We live in a closed and regenerating life scheme. The recuperative and recycling powers of the Being which is this planet are far greater than the alarmists lead us to believe. I might just cite an example of what I mean. An eruption of a volcano creates a multitude more air pollution in one day than Los Angeles could probably create in 1,000 years, and yet the planet has survived the eruptions of volcanoes many, many times.

This is not to condone in any way the creation of pollutants, but merely to point out that air pollution is as much a natural phenomenon as it is a manmade phenomenon. And an understanding of nature's ability to cleanse the air would lead to a much greater understanding of how man can cleanse the air.

Keep in mind that the scientific answers to all of the problems of the day are there for the looking. They are not hidden; they are out in the open. It is merely the lack of understanding and science's inability to grasp the obvious which keeps these problems from being solved. I don't mean to sound simplistic, but the air pollution problem is a good example. The planet obviously has the capacity to recycle its air. The challenge to science is to understand this obvious and very apparent capacity. It would then be able to solve the problem of air pollution. The answers are all there for us to find.

Japikse: This strikes me as a very important point. Can the same be said about solving the energy crisis?

Einstein: Why, of course. The energy problem is simply a lack of oil. *[Laughter.]* Well, I'm serious. If you look at the problem in its most simple terms, it's that we don't have enough oil. But the planet as a whole obviously doesn't have an energy shortage; it's just us humans. The planet has existed for billions of years without burning oil and will continue to do so. There are other means and infinitely better ways to harness energy. The planet has other ways of generating energy, and science can discover those ways.

We are in a transition period right at the moment. We have been relying so much on one source of energy production that we are overreacting to the discovery that we won't be able to use that one source of energy indefinitely.

Leichtman: Of course, some people are going to read this evaluation of the situation and complain that you've just given us the energy equivalent of, "Let them eat cake." *[Laughter.]*

Japikse: Let them drink oil. *[More laughter.]*

Leichtman: Let them burn wood. Of course, you are absolutely right; we have had an obsessive concentration on oil as an irreplaceable resource, and that's rather foolish.

Japikse: Would you talk a bit more about ways of approaching the energy problem more constructively?

Einstein: Well, most of the approaches scientists are taking are basically correct. The problem lies more in the public and political perception of the shortage, and the tendency to limit the solution to current methods of generating energy. There are many ways to harness energy. There are many civilizations which have existed prior to our present one—or which exist now in other realms—and their means of generating energy

have often been quite different from what you are used to now.
Leichtman: Were these technical civilizations?
Einstein: Sure.
Leichtman: Is there a good probability that eventually scientists will be able to efficiently harness the power of the earth's magnetic field as a source of energy?
Einstein: For limited applications, yes. There are several ways to approach this problem of energy. One is to look at the physical realm and realize that there are various options for harnessing the physical sources of energy already known to mankind, be it oil—
Leichtman: Wind, rainfall, tides—
Einstein: The sun, magnetics, coal, gas, motion—
Leichtman: Hot air. *[Laughter.]*
Einstein: Yes, there will never be a shortage of that, will there? *[More laughter.]*

The other way to look at the problem is to realize that the ultimate energy source does not occur in the physical plane! Until this realization is commonly accepted, humanity will be limited merely to making the best use of what it has, physically.
Leichtman: Then nuclear fusion or fission—
Einstein: Are not the answer.
Leichtman: Right.
Einstein: Nuclear energy is still a physical form of energy. It is generated by doing something physical to matter. It uses physical matter to produce physical energy. In that sense, it is the equivalent to burning coal. It has its efficiencies and its inefficiencies; it has its specific applications, but is not suitable for other applications.

We're smart enough right now to know the limitations of the physical varieties of energy and how each type of energy should be used. But the real solution to the energy crisis will come with the realization that the ultimate source of energy is a nonphysical source.

This realization is many years off.
Leichtman: I take it that when this insight arrives it will

not just be a philosophical realization but a practical development which we can use to operate our refrigerators and automobiles. It will be a tangible force.

Einstein: Oh, of course. There are some scientists who are working in the right direction even now. They have had a glimpse of the possibility. But as with most major breakthroughs, it's going to take many, many years to achieve.

Japikse: When you say "many, many years," are we talking several generations, or thousands of years?

Einstein: It will be several generations until it is an efficient and usable source. This doesn't mean that there aren't some people who have the right idea, because there are. They are going in the right directions. But it will be many years until this nonphysical source of energy is harnessed practically.

I might add that this form of energy has already been harnessed previously in the history of humanity. So the challenge to science is a problem of recollection, rather than a problem of making a new discovery.

Leichtman: Atlantean science returns! *[Laughter.]*

Japikse: The future lies in the past. *[More laughter.]*

Leichtman: I presume this breakthrough will be coordinated to occur in harmony with the evolution of humanity and civilization.

Einstein: That is a major problem. If the answer were to arrive prematurely, it could be very harmful to civilization.

Japikse: It was certainly harmful to Atlantis.

Leichtman: I suppose the situation is similar to the one Tesla faced, with some of his inventions. If they had fallen into the hands of international terrorists, the consequences to civilization would have been disastrous.

Einstein: It would be like giving a gun to a three year-old. Well, this is a valid point. Certain technological breakthroughs only become possible when we have a satisfactory technological base and the ability of humanity to accept them and use them wisely. When this nonphysical energy source is tapped, it will become extremely easy for virtually anyone to generate enough energy to blow up the planet. That would be very dangerous, until humanity has grown up a bit more than it has.

Leichtman: Of course, we are almost at that point now. Our stockpile of atomic and hydrogen bombs is large enough to kill everyone on the planet 20 times over, I think.

Einstein: This is true.

Leichtman: Talk about a wretched excess! Who needs enough power to kill everyone 20 times?

Einstein: Well, I spoke out about this danger many times. I don't think it is necessary to spend much time here reiterating what I have said in the past.

Leichtman: No.

Japikse: But it is relevant for people to be reminded of the potential dangers of scientific breakthroughs. As we look for more and more powerful answers to our problems, we have a responsibility to grow up and become more civilized.

Leichtman: Yes, I think these comments certainly apply to the theme of this project—the work of the priest of civilization, the priest of God, whether it is a writer, author, doctor, or scientist. The priest of God certainly has a responsibility to civilization to make sure his work is helpful, not harmful.

Can we talk some about the scientist as a humanitarian, and the responsibilities of the scientist in terms of his training, his outer work, and his relationship to people outside the field of science?

Einstein: Of course. It should be clear that many times, for good reasons and bad reasons, scientists as well as others get caught up in the mere physical phenomena of life—in finding solutions to problems and making major breakthroughs just for the sake of discovery, rather than because the discovery advances a basic human purpose. Scientific work sometimes becomes a goal in itself, justified in the holy name of "pure" science. The classic example of this is what we've just been talking about—the scientist who harnesses a new kind of energy, thereby unleashing forces which should not be given to those who lack the enlightenment to use them wisely.

There's much science fiction written about this particular problem.
Leichtman: Yes.
Einstein: I know the pressures that can be involved. The atomic and hydrogen bombs were all developed in the spirit of so-called pure science and given to a world which was not prepared to use this science wisely. These breakthroughs probably should never have been disclosed.

A lot of pure science is done in the name of hate and warlike attitudes between nations. The arms build up today among the major countries is one of the great incentives for new technological advances, and yet the ultimate goal is mere military advantage and a greater capacity to destroy. As with any endeavor, science can be perverted to the ends of those who wish to do harm or evil to the world, rather than make the world a better place. So it's a harmful illusion for the scientist to become wrapped up in the "science for the sake of science" argument.

The reasoning goes something like this: "I am the brilliant scientist who can make these major breakthroughs. What the world does with them is not my problem." Well, what the world does with them *is* his problem. But it is easy to rationalize the problem away, because the pressures can be great.

Japikse: Exactly what are the pressures?

Einstein: One of them is the loss of scientific prestige by withholding or not disclosing a breakthrough. There is an intense desire among scientists to be well-respected and well-known in the scientific field. It is very difficult not to publish new information in the name of humanity.

Of course, scientists aren't the only ones who seek respect and fame in their worlds. Most creative people do. And certainly politicians do. But the scientist has a unique problem in that many of the fruits of scientific labor can be used for physical destruction, which is not true in other fields of creativity. I guess politicians are good at using the fruits of scientific labor to cause damage, but that is precisely why the scientist has a special responsibility.

Leichtman: Who's holding the gun to the heads of scientists forcing them to create inventions which destroy life and property?

Einstein: That's my point: the pressures are far more subtle. Usually, there is the hope or promise of some kind of gain. It's not always just prestige, however; sometimes a scientist is caught up in political rhetoric. He begins to believe that there are good reasons for his country to develop an advantage in destructive capability.

Leichtman: And of course we have in this country an enormous military industrial complex which seems to pay scientists very well, in addition to supporting their line of research and invention. Well, what are the solutions to these problems? Better education for scientists? Some kind of humanitarian alliance among scientists?

Einstein: That's an interesting question. I suppose everyone who has participated in these interviews has grappled with this basic question in terms of their field of enterprise. When you discuss the problems of society and try to come up with solutions, the all-encompassing answer always is that humanity as a whole must become more aware and must learn more about the purposes and principles of human living. Mankind needs to learn to put principles above personal gain, individually and collectively. Of course, this is a very starry-eyed statement, but until people start taking it seriously, none of the other possibilities will stand much of a chance of succeeding.

Both of the possibilities you mentioned are good ones. The scientific community should become more compassionate, more protective of humanity, and more concerned about the impact of its discoveries. It should realize that there is more to life than scientific prestige, and if releasing a discovery or invention will lead to increased destruction, it is better that it be suppressed. But it is not just enough for scientists to endorse these principles, even though it would be a good beginning. The whole issue revolves around the wisdom and intelligence and compassion of the world in general. And these qualities are conspicuously lacking in a few pockets here and there.

Leichtman: Are there specific recommendations you can make for improving the education and training of scientists, so that it emphasizes these humanitarian principles more?

Einstein: My specific recommendations would pretty much follow the ideas we've been talking about this afternoon. There ought to be instruction which would help scientists appreciate that the full scope of their inquiry must embrace a larger perspective than just physical observation and study. But the problem with trying to teach this very fundamental principle is that it gets tinged with religion and all the preconceptions associated with it. It's very difficult in our modern world to teach a proper evolutionary and metaphysical background for science. Using nonscientific terms such as "God" and "creation" is very difficult; the discussion soon becomes warped and politicized.

Japikse: Yes, you were well known to have strong mystical tendencies in your approach to science, but I find very few scientists who are comfortable talking about that aspect of your life.

Einstein: The ideal would be to give scientists some background in the subtle forces of life, a knowledge of how they behave, and an appreciation for the different levels and planes of existence. If this could be taught without associating these concepts with religious assumptions and creeds, it would be very beneficial.

In addition, I would like to see scientists trained in the ethics of working on behalf of humanity. They need to be given a framework whereby judgments can be made ethically regarding the propriety of disclosing potentially dangerous information.

Leichtman: You seem to be presuming that every scientist would always know exactly when he is about to open a Pandora's box. Is that always possible?

Einstein: No, you are absolutely right. It is often a very difficult thing to know.

Leichtman: A good example of this difficulty, it would

seem to me, is the genetic manipulation currently being done. It is not exactly clear how harmful this could potentially be. The hope is that it will be highly beneficial—but horrible things could come from it, too.

Einstein: I'm not advocating that scientists shy away from pursuing new discoveries, simply because they may be harmful. What I am saying is that better judgments must be made, once a discovery occurs, as to whether humanity is prepared to use it—and its implications—in an enlightened way.

You would be surprised, when you look at it from this side, how many inventions and discoveries have *not* been disclosed by scientists, precisely because humanity was not prepared to use them wisely.

Leichtman: I was about to ask you about that. It does seem that some scientists do make these judgments.

Einstein: Absolutely. There have been some wonderful inventions and discoveries which would have been quite an advancement for the planet as a whole but also had so much power for destruction, if misused, that the scientists involved decided, with a lot of prodding from the inner side, that it would be better not to make them known.

Leichtman: Yes, I've run into people who have made that kind of decision. They were all individual, personal decisions, though. This was never a case of someone being hounded by others into suppressing his discovery.

Einstein: Right. It's a little more difficult to make that kind of decision if the scientist is working in a government sponsored program, however. A bureaucrat is bound to run in, grab it, and steal it.

Leichtman: Could you give us an example of some of the discoveries and inventions which have been suppressed?

Einstein: Well, because of the energy crisis, there have been numerous discoveries in the last several years for harnessing energy. Some amazing prototypes have been developed. The original motivation for working on these devices has usually been to light our cities and power our automobiles more efficiently. But when the prototype has actually been developed and the scientist put his "evil hat" on, he has begun to think about what would happen if the president of a banana republic somewhere would start to use these devices. So he has decided that humanity would be better off without his invention at this time.

Leichtman: Yes, I know someone who scrapped some electronic devices which could manipulate consciousness to a surprising extent. Well, it sounds as though the project of getting scientists to think in these terms is basically one of moral and philosophical ethics.

Einstein: I might add that it applies not just to the work of the physicists and the chemists, but also to the work of such scientists as psychologists and sociologists. They have exactly the same responsibility to make sure that the ideas and projects they develop truly serve the public, and not harm them.

Leichtman: What kind of potential harm could the ideas of a psychologist or sociologist do?

Einstein: It's often quite subtle, and not everyone has the capacity to recognize it. There are certain brainwashing tactics, for example, which are now being incorporated into psychological practices. Of course, the psychologists get very annoyed when someone accuses them of brainwashing, but the use of the term is valid.

Like physicists and engineers, the psychologists tend only to see the positive side of their theories and how they help people, but some of these practices are very dangerous. There is one new application of hypnosis in particular which is rapidly becoming popular in America that is a veritable time bomb threatening the unsuspecting public. There are also the bizarre theories and new techniques for managing "sexual problems" which are now being used by some unorthodox schools of psychology. The potential harm of these new developments in psychology—psychological technology, if you will—is going to be difficult to assess, but some of it will be quite detrimental. It will also be hard to reverse the damage, once it is done.

The influence of new discoveries in sociology is more

indirect. Sociologists usually don't have much direct contact with the public, but they do influence politicians, political scientists, and other social planners. Some of their discoveries about the reasons why certain groups exhibit antisocial behavior are valid, but superficial. Just like other scientists, they often are looking only at physical phenomena, and do not embrace the larger perspective. They do not look at the physical plane from the next higher level. And the solutions they frequently propose, as a result of their research, often actually deepen the real problem. In particular, there are a lot of strange ideas coming out of the sociologists who study the so-called "disadvantaged" or the "victims" of society. They are not seeing the whole picture, and if society accepts their conclusions, the damage done could be very serious.

I don't pretend to be an expert in these fields. But I do think it is important for people to understand that it is not just the hard sciences which need to improve their ethics and be aware of their potential for harm. Anyone who would serve humanity must struggle with the possibility that he or she might unwittingly harm humanity instead—and take action to make sure this does not occur. It is part of the responsibility which falls to those who would serve.

Japikse: You mean it's not easy being a priest of God? *[Laughter.]*

Einstein: It's certainly no role for anyone who blithely assumes that we live in the best of all possible worlds. *[More laughter.]*

Leichtman: I should think your comments would also apply to the more militant religious groups. Some of the theories and techniques they embrace are certainly dangerous.

Einstein: And militant political groups, too. Yes, but they aren't often confused for scientists. *[Laughter.]*

Keep in mind that *wherever* there is power to be wielded by a group, there is also the risk of misusing that power. My basic point is that even though the discoveries of researchers in all scientific disciplines may be valid and useful in a limited way, not all of them will be useful to society in the long run. Some may be positively harmful—and this should be a concern for all researchers.

We are moving into a period when it will be more and more important to make these decisions. We are going to have to examine philosophically the *purposes* of psychology, sociology, and the hard sciences to determine how well they are being served. Do our sciences *serve* humanity and civilization? This is a fundamental question. For years, we have been dissecting the mind and the atoms and gazing at the stars and amassing an amazing amount of information. Now it is time to stand back a bit and look at why we are gathering all this scientific information. I'm not saying it is wrong to gather data—not at all. I'm simply suggesting that we need to better understand why we are doing it—and the benefit humanity will receive from proceeding along any particular scientific line of inquiry.

There are many advances we can make in this regard.

Leichtman: For a start, it would help just to get scientists to appreciate the full significance of the fact that we all live on the same planet, are all part of the same race, and all share the same problems.

Einstein: Yes.

Leichtman: Is there other advice you can give on the education of scientists?

Einstein: I have to say it is a mistake to divorce the spiritual side of human life from the scientific side. We do not integrate these two sides of life very well. The mode of the day seems to be to go worship God, and when God is done with us—or we are done with God, which is closer to the truth—then we go do our scientific work. There is very little bridging or understanding between the two. In fact, there is almost a kind of elitist attitude on the part of scientists; their religion, in a sense, becomes either their inventions and discoveries or the scientific methodology they follow.

So I want to stress the usefulness of educating the scientist about the spiritual side of life—or, as our friend over here

[C.W. Leadbeater, whose interview appears in *The Inner Side of Life*] calls it, the hidden side of things. We need to give the scientist the perspective that he is merely *in the process of discovering what God has already thought of and created*. True science is the discovery of divine vision—and the understanding of how this divine vision has manifested itself in the physical plane. It is not and never should be the simple discovery of some isolated little phenomenon.

The scientist's education should include the concept that there is order in the universe and that an intelligent force has, in fact, thought through and given life to the phenomena science observes. This needs to be a part of science. It would be an important leap forward.

Leichtman: Of course, you realize that many scientists, particularly the ones who control the education of scientists, would look askance at such an idea. Many of them automatically—and I might even say unthinkingly—consider religion as something which only undermines and pollutes the pragmatic basis of scientific endeavor.

Einstein: And that's precisely the difficulty, as I mentioned earlier—that the teaching of the value of the hidden side of life becomes warped by religion. I'm not really advocating a religious approach to the hidden side of life, but rather a scientific or metaphysical approach. You can't teach a scientist devotion. It is sort of a contradiction in terms.

Leichtman: Well, they can develop a respect for the truth, and that would be a reasonable equivalent of devotion.

Einstein: Exactly. The scientist needs to include God in his scheme of things, but he must approach God on his own terms, in his own way—through the path of inquiry and respect for truth, not through religious devotion.

Leichtman: The path which leads to the wisdom aspect of God rather than the love aspect. Yes, I think that concept could be incorporated into the scientific method.

Einstein: It could be done, and ought to be done. But it would be very difficult, precisely because of the people who staff the universities and control the education of scientists. But it isn't impossible.

Leichtman: That begs the question—

Einstein: We've begged a lot of questions today. *[Laughter.]*

Leichtman: Well, I'm just thinking that while science needs a greater measure of spirituality, it is also true that religion needs a greater measure of scientific pragmatism.

Einstein: That would be for someone else to discuss, though. Why don't you go ahead. I'll give you five minutes. *[Laughter.]*

Leichtman: It is an interesting corollary, don't you think?

Japikse: Something like East meets West, but in this case Science meets Religion.

Leichtman: It certainly would be an enrichment of religion to get rid of some of its wool and bombast and replace it with intelligent thinking.

Einstein: Unfortunately, the people who control religious education are just as rigid and skeptical of this blending as the people who control scientific education.

Leichtman: I suppose in the case of science, a few enlightened members of the scientific community here and there will start to set an example and articulate the ethics and philosophy you've just been describing, to their peers, and it will slowly take root and evolve. I assume there is substantial interest "upstairs" among you spooks to promote the spiritualization of science—or at least a new definition of humanitarian ethics among scientists.

Einstein: Absolutely. There's quite a push on, because we are getting to a very critical period in the development of technology and science. As you know, the last time we reached this level of ability, we did end up blowing ourselves up.

Leichtman: Yes.

Japikse: Should there be an explanation of what that's a reference to, for the benefit of those readers who missed it in history class? *[Laughter.]*

Leichtman: Yes, go ahead.

Einstein: Why don't one of you explain it?

Leichtman: Well, it's a reference to the final days of Atlantis, approximately 14,000 years ago. Atlantis was a civilization centered on a large island in the middle of what is now the Atlantic Ocean. The history of this civilization has been lost except for a few lingering myths, but it is said to have been quite sophisticated and technically advanced. The scientists of that era invented extremely powerful sources of energy and fantastic devices which transmuted matter and manipulated consciousness. In the end, the power sources were misused in such a way that it led directly and indirectly to the disappearance of the whole continent, plunging the Atlantean civilization to its death. Humanity then forgot about all this vast technology and scientific achievement and proceeded to slide into a very long dark age.

Einstein: We certainly don't want anything such as that to happen again. None of us is expecting it, at least. And I think there are signs that the more concerned members of humanity are taking steps to nudge civilization and science—as well as government—toward safer and more humane levels of operation.

Leichtman: Yes, I know what you're talking about. Beyond some of the bombast and strident rhetoric of a few of the environmentalists, there are many good people who are truly concerned about protecting our environment. These people certainly have a humanitarian motive and interest. And there is also a group of physicians who speak out very articulately about the possible dangers of radioactive poisoning and nuclear holocausts. I assume these would be examples of the point you've been making.

Einstein: Yes.

Leichtman: To change the topic to something far from the mainstream of orthodox science, it seems to me that you are hinting that the new frontiers of science are moving toward magic. If science is to take up the study of the subtle planes of matter and the forces and phenomena of those levels, isn't this really the study of true magic? I know this label offends many scientists, who consider it scientific heresy even to mention the subject, but what about it?

Einstein: In a sense, it is magic, yes. We are coming up to the exhaustion, so to speak, of the possibilities for analyzing physical phenomena. Science has been taking a magnifying glass and magnifying everything in the physical level of life to such an extent that there isn't that much left to look at anymore. So it will have to begin to appreciate the causes underlying the phenomena it has been viewing and analyzing—and attempt to manipulate these causes rather than just monitor the phenomena. Yes, that could be called magic—not sleight of hand, of course, but real magic.

The training of the scientist in the 21st century will therefore be quite different than now.

Leichtman: Will this transition to real magic begin to occur in the foreseeable future?

Einstein: No. We are still at a point where we are far too selfish and greedy. The value of science for most people is in using it to get ahead of their neighbors, to build the best weapons of destruction, or to grab the power which will make them king of the world. We have too much of this kind of thinking. The consciousness of humanity is no where ready for the type of exploration you are suggesting. But it is a nice ideal to think about and strive toward. It will be possible one day.

If I may, I'd like to add a few more comments about education—not the education of the scientist, but education in general.

Leichtman: Of course.

Einstein: I would like more people to realize that education should not just be limited to academic study. If an in-depth analysis is ever made of education, it will become clear that there is a time and a place for academic study but also a time and a place to concentrate on other important aspects of our human skills and talents.

Let me ask you to think of formal human education as

a process which should last from childhood to age thirty. In that span of time, there are various stages of development that the human child passes through, as he grows and learns. Each of these stages should be taken care of at the appropriate age. I will discuss each of these stages briefly, and then come back and make some comments about the meaning of education itself.

Obviously, one of the first stages in the development of the human being is learning the lessons of physical coordination. I would say that up to the age of ten years, learning physical coordination is the most important task to concentrate on.

From ages ten to twenty, the main thrust of development should be in the emotional area, learning the lessons of emotional stability and maturity. This may seem like an oversimplification to some, but it is interesting to speculate on.

From twenty to thirty, mental perception and intellectual discernment would be the major emphasis of development. Again, I am simplifying the analysis in order to make my point.

Now, it is interesting that what conventional education does is lump all of these together, from the time the child enters school until he or she is eighteen—or twenty-two if he goes to college and maybe thirty if he goes to graduate school and beyond. I don't think this is sensible. It has always been my conviction that we do not concentrate on the critical areas of development in the right years. In fact, we leave out the whole issue of emotional development and concentrate on physical and intellectual development—and even this is done in the wrong years!

Three facts are essential in developing a sensible format for education. First, mature emotions are a prerequisite for correct and intelligent thinking. If we do not have mature emotions, then our feelings will tend to interfere quite substantially with our thinking and our intellectual pursuits.

Second, the growth of the intellect occurs in the latter part of our development; certainly, we are in the late teens before this intellectual ability really begins to flower. And yet, this is precisely the point where most general education usually stops!

Third, most of the emotional development which occurs between the ages of ten and twenty goes relatively uninstructed. There are virtually no courses and no instruction in right emotional development and behavior. And so, for the average person, the educational process stops right when it should be beginning. We need to rethink our philosophy and educational methodology, so that we can identify the real educational needs of the growing person and tailor those needs to the developmental cycles of the human being as a whole. This is not to say that certain basic educational tools cannot be taught at an early age. Nor do I think that nothing should be taught between the ages of zero and ten except athletic abilities. Certainly, all of the basic lessons of mathematics and reading need to be taught in these early stages, but the primary emphasis should be on the lessons of physical coordination, skills, and health.

Japikse: Language is a good example of what you are saying. If a young child is taught a foreign language, he can pick up the pronunciation very quickly and very accurately. But a teenager has a much harder time, because the physical patterns of manipulating the mouth have become set. Yet a young child has no capacity to comprehend the logic of grammar, and even a teenager has difficulty doing so. It is in the twenties when mature understanding of any subject becomes possible.

Einstein: Exactly. I hope people will see that the stages I have described are not rigid divisions, but focal points of development.

Leichtman: Is this program feasible, though? Many people would want to marry and start their family before age thirty. Yet you would have them still in school learning their professions and not being free to launch their careers or even their personal lives.

Einstein: I'm not sure that's the case. Much of the intense emotional development which goes on in the teenage years is left untouched by our educational system. It's very interesting to me how it is totally ignored educationally.

Leichtman: I think it's been getting more attention in the last twenty years—certainly in America, at least.

Einstein: In my case, I had virtually no interest at all in facts during my teen years. I was more interested in what was going on around me emotionally. I had very little incentive to study facts, even in the sciences. And when I did, because I had to, I plodded through it.

Of course, when I reached the point where I became interested in the intellectual elements of life, I was ready and capable and genuinely interested in learning about facts, and I threw myself right into the study of them. But the educational process didn't help me discover this interest. It grew from within me. And I think it is sad that our educational systems do not recognize these natural stages in development.

To answer your question, not everyone would have the interest to pursue intense intellectual development that I did—or the two of you did. They would go off and start their families and careers. It would be expecting too much to demand that they stay in school through the age of thirty.

Leichtman: Well, I know many who would not want to pursue mature emotional development, either.

Einstein: Well, they should be forced. *[Laughter.]*

Japikse: If not for their own sake, then for ours. *[More laughter.]*

Leichtman: I agree, but this statement will drive certain psychologists right up the wall.

Einstein: Good. That's where they belong. *[Laughter.]*

Well, as I said, I kept my original statements about the stages of human development deliberately simplistic, so they would be easier to comprehend. It might very well be intelligent to offer two educational roads. One would be the general education for the person who did not want to go on to intense intellectual pursuits. The other would be a specialized education involving a far more intense training, first of the emotions and then intellectually.

Leichtman: Of course, many parents are very ambitious for

their children. They know their little five year-old is destined to be a doctor or a scientist, and they will push the intellectual development almost irrationally.

Einstein: Which is one reason why we need to better understand the cycles of human development.

Leichtman: Well, how about the children who demonstrate precocious talents at a very young age. Shouldn't those talents be honored at that time?

Einstein: Sure, but what if they weren't? What if they spent their time becoming emotionally mature? Wouldn't that be a satisfactory outcome?

Leichtman: It certainly would be different than the norm. *[Laughter.]*

Einstein: I make no excuses for myself in this regard. I suffered emotionally for many years, and it affected my ability to think and my desire to pursue intellectual studies. It would have been a great advantage to me to have received training in an educational setting which would have helped me deal with my emotions in a more disciplined fashion.

Leichtman: Who would make the decision as to whether a child would pursue the intense intellectual education or the more general education?

Einstein: The individual would.

Leichtman: Do you think kids would be allowed to make that decision? Do you see ambitious parents saying, "Okay, son, it's all up to you. Go ahead and be a janitor. We'll support you all the way." *[Laughter.]*

Einstein: I don't think you're getting my point. Ambitious parents will distort any educational system, no matter how enlightened. My point is that our present educational system is not geared to take advantage of the natural cycles of human development—and it ought to be.

I am not proposing that all intellectual studies should be eliminated during the teenage years. I am merely stating that the teen years are the natural time when a child is focusing intensely on emotional development, and the primary emphasis of education at the time should be to help children become emotionally mature—to learn the principles of right human relationships, for example, and techniques for disciplining the emotions effectively. After all, these are lessons the two of you find you have to teach adults—precisely because they didn't learn them when they were children.

Leichtman: Do we have the psychological theories and techniques—and the teachers—to teach emotional maturity on a large scale in our high schools?

Einstein: Of course not.

Leichtman: I don't think we have the proper role models, let alone the teachers and techniques.

Japikse: Yes, there is the danger that we could end up with instruction on how to be a better con artist or manipulator.

Leichtman: If you look at the heroes of the high school set, they are usually rock and roll stars, athletes, and people who become well-known because of the glamour and flair of their life—not because of their maturity or competence or contribution to society.

Einstein: Which is precisely why we need to rethink our educational priorities. The current system is doing a terrible job of guiding our young people toward emotional maturity.

Leichtman: Yes, I think you've hit on something important here. There would be immense value in the kind of program you are proposing. I see many people whose brilliant mental capacities are dissipated in excessive attention to their frustrations, a bad self-image, and their fears. It is a tragedy to see this magnificent potential being so grossly under-utilized because of basic emotional immaturity. I suppose this is another issue where civilization must evolve a bit more before these good ideas can be adequately implemented.

Einstein: It may be that civilization will not evolve that much more until this type of philosophy and practice revolutionizes our educational system. It is the basic selfishness and materialistic greed of humanity which limits mankind's capacity to achieve major scientific breakthroughs. Making training in emotional and ethical maturity a major focus of our educational system would go a long way to answering a basic need of humanity. I think it would help a great deal. In fact, I *know* it would help.

I am not saying there would not be enormous problems in establishing this kind of educational system. I'm just suggesting that certain people ought to start thinking about pursuing the educational objectives I have outlined.

Leichtman: I think you are absolutely right.

Japikse: Amen.

Einstein: That should be enough to get the right people thinking. If I may, I would like to talk some more about nonplanetary issues.

Leichtman: You mean intergalactic Zionism or something like that? *[Laughter.]*

Einstein: No, no.

Leichtman: I've heard of the wandering Jew, but this would be too much. *[More laughter.]*

Einstein: I want to comment a bit more on the essential unity of the entire solar system. Contrary to popular belief, the space between the planets in the solar system is not empty. Subtle streams of matter firmly connect all the planets and, in fact, there is a constant interchange of matter on the subtle planes between the various planets.

Leichtman: That interchange is virtually absent on the dense physical level, though?

Einstein: Well, there are gravity pools and so on which make it apparent on the physical level, too. But the interplay on the subtler levels is much different from what we can find on the dense physical plane. It is much more extensive.

Understanding this basic concept opens up many possibilities. The best way to communicate with other planets, for example, would be on this subtler level. And travel, too—it would be much easier to travel to Venus on this subtler level. The vehicle you would travel in would be much simpler to produce, too.

Leichtman: You mean interplanetary travel is practical only when you can do it on the astral level?

Einstein: So it seems. Or preferably on an even more subtle level.

Leichtman: But that makes it impractical for those of us with dense physical bodies!

Einstein: That's your problem. *[Laughter.]*

Remember the vortex we talked about—the vortex which is created as a planetary being interacts with matter? On the subtler levels of matter, there isn't as great a demarcation between the vortex of one planet and the vortex of another as there is on the physical plane. It is more a vortex of subtle matter which is moving continuously throughout the solar system and, in fact, throughout the universe as a whole. The particles of matter tend to be more spread apart the further you move away from the center of the vortex, but they are still responsive to the center. This makes it easier to travel between and have interplay between the astral vortexes or astral bodies of planets than between the physical forms of the planets. Let's take astral matter, to focus on just one of the subtle levels for awhile, and picture the sun as the center, with each of the planets being a vortex of astral matter. Each vortex could be visualized as a whirlpool. This matter not only swirls around the vortex, but as it gains momentum, some of it is thrust away from the gravitational force of the vortex and out into space. It would then travel through space, attaching itself eventually to another vortex, where it would swirl around for awhile, as it did before.

Leichtman: Are you describing a combination of centrifugal and centripetal forces cycling back and forth?

Einstein: In a sense, but this is happening on a subtle level, so it wouldn't appear in exactly those ways.

Japikse: Are these vortexes directly associated one to one with the planetary beings?

Einstein: Yes.

Japikse: Does the matter formed as a vortex exist before the planetary entity or Idea is born?

Einstein: No. The vortex is formed at the time the planetary being comes into manifestation in matter.

Japikse: Then it is after this point that there is this interaction of matter from one vortex to another.

Einstein: Yes. For instance, if you were to pick one particle of matter and follow it for a period of millions of years, you would probably follow it from vortex to vortex. Sometimes, this particle of astral matter might be traveling between vortexes hundreds of thousands or even millions of years; at other times, it might be in a vortex for a similar length of time. But all of this matter is constantly moving. And this movement has an effect on the nature of the matter, too. In one vortex, the matter might become more dense, and then when it has matured and become less dense, it would move on to another vortex. Do keep in mind, however, that matter is only a mechanism which provides the vehicle for the movement of consciousness, whether it is the vast consciousness of a planetary being or just the consciousness of an individual human being.

Japikse: Is this movement random, or is it under the control of the ensouling consciousness?

Einstein: It is generally random. Also keep in mind that we have been talking about astral matter, just one type of matter, for the purposes of our discussion here.

Leichtman: Do you mean that this movement of astral matter is random regardless of it being in the body of an elephant or the body of a whole planet?

Einstein: Oh, I see what you mean. There is a distinction between the astral matter which makes up the emotional body of an entity and the matter which makes up the remainder of the astral plane. The astral matter which is attracted to an individual entity is relatively fixed. The rest is in a far more random motion, and this random motion is particularly obvious in the space between vortexes—that is, between planets.

Leichtman: Is there some correlation here with the phenomena of astrology? Or is this explanation you have given us the scientific basis for astrology?

Einstein: Oh, there is a distinct correlation, yes. It is a bit difficult for me to discuss this whole subject thoroughly, because the interplay takes place on a subtle dimension with laws and activities which are quite different from the laws and activities of the dense physical plane. Motion is different, for instance. It is not always in a nice, neat straight or curved line. You can have matter expanding and contracting at the same time. Matter also has "enhanced" magnetic properties at these subtle levels. I can see it all very clearly now, and understand it quite well, but it takes time to communicate it fully to people who must necessarily use physical terms and concepts to understand it.

Leichtman: I appreciate what you are saying. I didn't mean to imply that we expect neat and simple answers.

Einstein: Well, let me try to explain it as best as I can. Subtle matter, especially astral matter, is very magnetic. Movement at this level is relatively fluid, compared to the dense physical plane. There are forms, but they are mercurial. They tend to pulsate, and movement can be in more than one direction at the same time. It is, after all, another dimension of existence and must be understood on its own terms.

Now, if you consider that matter is the substance an ensouling consciousness uses for self-expression and that matter has magnetic properties, you begin to establish the basis for astrology. Since this subtle matter fills the space between the planets or vortexes, it means that our astral bodies can rather easily be influenced by the magnetic matter of other planets.

Leichtman: Do you mean that our astral bodies pick up signals from other planets—something like picking up radio waves?

Einstein: No, it's not quite that way. It's more like having a container filled with a solution of iron sulfate. Such a solution would contain billions of iron ions which would respond to magnetic influences in near proximity.

The outer planets act somewhat like giant magnets affecting the magnetic properties of our astral bodies. Of course, the type of magnetism is different, but the analogy may help you understand how astrological forces influence us.

Leichtman: What accounts for the apparent changes in astrological conditions and influences throughout the year? And why aren't they the same for everyone?

Einstein: I'm not sure I can give you an adequate explanation. I understand it, but again we run into the difficulty of explaining phenomena which do not have a good correspondence on the physical plane. For instance, how do you explain something which is both fixed and fluid at the same time? Things like this do happen on the subtle planes where astrological forces have their impact.

Japikse: Can we at least say that astrological forces are the total influence of planets and other heavenly bodies—that is, the spiritual, mental, and emotional influences of these vast beings? When I try to explain the real meaning of astrology, I find people are stuck on thinking of the planets in only their dense physical forms, which does make the whole concept of astrology sound ridiculous. It's preposterous to think of psychological influences emanating from dense physical balls of dirt. But it's not preposterous at all to think of these influences emanating from astral, mental, and spiritual vortexes.

Leichtman: And at the other end of the scale, there are the "true believers" who think they don't need any explanation at all about how these forces work—and wonder why we bother.

Einstein: These are useful points. I am not an expert on astrology, and I really cannot give anything other than some general ideas and explanations on the subject. But I do want to make it clear that planets, like human beings, have an indwelling life essence or soul, and that they use several types of matter for their manifestation in the various levels of life. We all know that individual humans radiate their thoughts and attitudes to a greater or lesser extent—and planetary beings do the same on a much grander and more powerful scale. *This* is the basis for astrological forces. So you are right, Carl—they are the mental and emotional influences of vast beings, planetary and solar.

What I would like to stress here is that these influences

travel from vortex to vortex through the medium of subtle matter, much as electricity moves through a copper wire. Except for this continuous magnetic link in subtle matter, much of the astrological influences would not reach our personalities which, after all, are made up of subtle matter, too.

Leichtman: How would you explain the fact that the same planetary positions affect people differently?

Einstein: I'll try to answer that question, but please remember that I am neither an astrologer nor an expert on the anatomy of your subtle bodies.

At the time of your birth, the relationship of all the major subtle magnetic influences is imprinted on your newly-created astral body. This fixes your receptivity to the magnetic forces of the planets and other astrological sources. It is somewhat analogous to creating an orthodox magnet out of a bar of steel. You fix the two poles of the magnet; one end will always be the north pole and the other end will always be the south pole, for as long as the steel bar is a magnet. For humans, the situation is vastly more complex. You are dealing with a moderately fluid body instead of a neat bar of steel, and you are creating more than a dozen "magnets" instead of just one. To make matters even more complex, each of these dozen or so magnets is located in a different part of the entire body of the emotions. And each is responsive to only one vortex or planetary entity, depending on the astrological conditions at the time of birth.

The magnetic responsiveness of your astral body is therefore relatively fixed. But the sources of these magnetic influences—the vortexes—are not fixed. They move about. As a result, their influences vary greatly throughout long cycles of time, for humanity in general and for each individual in particular. As the vortexes move throughout the solar system, it produces a direct effect on the astral matter of the individual here on earth, causing its configuration to change.

Leichtman: That sounds very complicated.

Einstein: Complicated, yes—but hocus pocus, no. The effect of a particular planet is conveyed through a stream of subtle matter between itself and the individual it is affecting. It is the subtle matter which provides the vehicle for astrological influences. This is a fact which is very clear to me.

I suppose I have just proposed a set of ideas which will be considered scientific heresy in some circles, but many people have a lot to learn about matter and the solar system.

Is that enough on astrology?

Leichtman: Yes. I'm eager to ask something about the transmutation of physical matter. I'm referring to the process of turning one element into another, like transmuting lead into gold or garbage into gasoline. Is this feasible on a commercial scale?

Einstein: To transmute matter from one element to another?

Leichtman: Yes.

Einstein: Of course it's possible. It is not feasible today, given our current technology and the limited scope of scientific thinking. The modern scientific perspective on matter limits the activities of science to a level where the practical transmutation of matter is not possible.

Leichtman: Would this process, when developed, take enormous amounts of energy to operate?

Einstein: It would take huge amounts of energy, but the energy needed will be available by the time science is ready to discover practical ways of transmuting matter.

Leichtman: There are a number of people who are working with little devices they claim will transmute matter—the De la Warr box and the Hieronymus device. Not all of the people working with these devices make this claim, but some do.

Einstein: I am not aware of this, but it is certainly possible to a very limited and crude degree.

Leichtman: There are other people who claim that by making friends with hordes of elves, fairies, gnomes, and angels, they can intrigue them to transmute matter.

Einstein: Well, that's not true.

Leichtman: I didn't think so. It seemed to me they were producing an acceleration in the natural processes of decay, rather than the transmutation of matter.

Darn, I thought we had the solutions to the pollution problem here! *[Laughter.]*

Einstein: Perhaps you do have a clue or two in what you just said.

Leichtman: Ahh. I'll have to think about that.

I would like to go back and ask a few more questions about the etheric plane and the scientific investigation of it. Since this is really the subtle aspect of the physical plane, it seems ripe for investigation at this time. In fact, there are some who predict that the etheric plane will be accepted as a legitimate area of research and that much will be understood about it in the next one hundred years. What can you say about this?

Einstein: You must remember that one of the first modes used in investigating the dense physical plane was to examine it with a magnifying glass. Some of the investigations into the etheric level of matter will begin in about as crude a manner. The first problem science will confront is how to view the phenomena of the etheric plane, which is invisible.

There is no question that there is much to be learned from studying the etheric plane. It will take science nearer to the cause of dense physical phenomena. The danger will come when someone decides, "Since we can see energy going from point A to point B, let's try to magnify it and change its quality, and then point it at someone and see what happens." Of course, the time when these dangers will be a real problem is in the future.

Leichtman: Meaning how long?

Einstein: It's closer than you think. I would say it would be close to the middle of next century.

Leichtman: Is it possible to explore etheric matter in terms of the human body, and wouldn't that be safer? There are a number of systems of diagnosis and healing which are based on working with the etheric body, even if they are not recognized as such.

Einstein: Oh, in medical terms, yes. There could be research done to monitor the effect of changes in the etheric body and how these changes affect the health of the dense physical body connected with the etheric body. But then you will have some guy come along who will decide to develop an etheric ray gun for the purpose of disrupting the etheric body and thereby killing the dense physical.

Japikse: Which would lead to a whole new cliché: "Etheric ray guns don't kill people; people kill people." *[Laughter.]*

Leichtman: Perhaps the more relevant idea for medicine would be to become more aware of the relationship between our individual consciousness and the matter which constitutes our subtle bodies. If we understand this relationship, we will comprehend why we can't just take five seconds to think saintly thoughts and expect it to lead to much of a change in ourselves. There is a certain inertia which overwhelms us once we take on an incarnation and have physical and etheric bodies. Once we are encased in matter, the range of our emotions and ability to think is limited, at least until the substance of our subtle bodies is purified, disciplined, and transformed.

Einstein: Yes, consciousness is limited by the quality of the matter it has to use for manifestation. Of course, matter is affected by the quality of consciousness, too. This makes for a complex situation of mutable and variable influences.

Leichtman: Will exploring these possibilities eventually help us to understand how the healing of the emotions can lead sometimes to a healing of the dense physical body?

Einstein: True emotional healing would involve, to some extent, the transfer of emotional matter out of a body and then replacing it with better matter. Once you change matter in one body, you can expect changes in corresponding bodies of the same system as well. This statement is a bit of an oversimplification, but as a generality, it is valid.

Leichtman: It sounds as though knowing more about this would help us better understand the very real phenomenon of spiritual healing through prayer. The invocation of invisible forces does bring about physical healing in some cases; you seem to be suggesting that physicists ought to go back to being alchemists, at least to a degree.

Einstein: It is more likely that medicine will take a different course, when it is well understood that many of the problems associated with the physical body can be linked to changes in subtle matter.

Leichtman: Maybe we will be able to get a Ph.D. in magic several generations from now.

Einstein: It won't be called "magic." It will be called "matter transformation."

There is one other point I would like to bring up and tantalize you with. We don't have enough time to talk about it much, but the subject is time itself. In effect, time is the movement of consciousness through matter. Since time involves matter, if there were no matter, there would be no time. Where consciousness is in its pure state, not existing in matter, there is no time. It does not exist.

Leichtman: Hmm.

Einstein: Time is a phenomenon which occurs only at the intersection of consciousness and matter. As consciousness moves through and transforms matter, it gives us the perception that there is a beginning and an ending.

Leichtman: When you use the term "consciousness," you are referring to the ensouling consciousness or spiritual essence, aren't you? Some people use the word "consciousness" to refer to any state without a physical body, such as the dream state or the life after death of the physical body.

Einstein: Yes, I was referring to the pure state of consciousness, the cosmic Idea not associated with any matter of any type.

Leichtman: I presume the phenomenon of time has a different meaning and different points of reference, depending on the type of matter involved—that is, the plane on which the events are occurring.

Einstein: This is perfectly true. We know that we can plan events in our thoughts but not perform them until later. In this case, some aspect of the event has already occurred in the plane of subtle thought matter. In fact, many of the dreams which contain images of future events come from this type of interception of a phenomenon which is already in progress or has already happened on a subtle level of matter.

On the other hand, not all images encountered in dream states are related to time. The human imagination is quite capable of conjuring up worries and fears which never come to pass. This is an entirely different issue, since most of the conjurations of the imagination are never intended to be grounded in the physical plane.

Leichtman: How do you tell the difference between simple imagination and events already in progress at a subtle level? I am sure there are a great many people who would enjoy a simple and neat answer to that question. *[Laughter.]*

Einstein: You seem to relish posing questions that have no "neat and simple" answers. *[More laughter.]* There is no simple way of separating out idle fantasy and daydreams from the shadow of real events to come. I can only make the theoretical statement that the phenomena of time are dependent on consciousness moving through matter. It will necessarily be a different phenomenon depending on where you view it—that is, the level of matter at which you are perceiving time.

I might also add that the activity of consciousness in matter is an essential ingredient of time. This could be the activity of large groups of people, such as a whole nation, or a single person. The problem is that fragments of thought can break off from the individuality of the thinker. These thought-forms are then capable of influencing the dreams and reveries of many people, but they do not contain the real power of the originating consciousness, so they usually do not manifest in the physical plane—for the secondary people perceiving them. They do not "come to pass." Perhaps that will help answer your question.

Leichtman: I'm not sure what the practical significance of this last observation is. I'm a bit lost.

Einstein: There may be no significance at all to many people. What I am saying is that time is relative to the plane of matter you use for viewing phenomena and the level on which the events you are viewing occur. As you penetrate into progressively more subtle matter, the phenomena of time are speeded up until, theoretically, they become infinite. Of course, this is another way of saying that the phenomena of time become zero when time is proceeding at an infinite speed, because the beginning and the ending are the same—in which case there would be no time. It's that simple.

Leichtman: I guess that means that our inner being has knowledge of how we are going to complete the rest of our life and what will happen to us. Are you slowly talking me into a case for predestination? I really don't accept that concept.

Einstein: Don't worry. For predestination to occur, you would have to assume that your whole personality would behave like a robot and that your whole private world was totally orderly and in harmony with divine plans. You would also have to assume that no human being ever had free will choices to make and that no goof ups, accidents, or disagreements ever occurred. Fortunately, this is not the case. Life would be so dull if predestination actually were the case that even God would get bored. I suspect He would pull out the plug.

[Laughter.]

Japikse: That bad, huh?

Einstein: Well, I'm being a bit facetious in assuming what God would do, of course. But in any case, it doesn't work that way.

Leichtman: Yes, my observations have always indicated that the inner being does have plans for the major events of a lifetime, but those plans often must be modified by the actual conditions the personality finds itself in, as well as daily interaction with others and life in general. Many plans never get implemented and many unscheduled events occur.

Einstein: That's part of the mystery of life. I think you are approximately correct in your interpretation of what happens to the intentions of the inner being. The same can be said, of course, for the intentions of large groups of humanity—whole nations, races, and the institutions of society.

Leichtman: Are we to infer that consciousness somehow benefits from its interaction with matter? You seem to imply, from these comments on time, that consciousness in its pure state is eternal. But that doesn't mean it is perfect or changeless, does it?

Einstein: You infer correctly.

Leichtman: I'm just thinking that if no real change or enrichment occurs in pure consciousness as it transits through matter, then creation is finished before it gets started.

Einstein: Well, I said earlier that it is consciousness that evolves, not matter. This is what I was referring to—the enrichment of consciousness through its involvement in matter.

Japikse: So even if we are not ending before we got started, we are at least ending on the same note as we started. *[Laughter.]*

Einstein: Coming full circle, as it were.

Leichtman: Well, do you have any final statement you would like to make?

Einstein: I would merely reemphasize that we need to modify the way we look at the phenomena of the physical plane. I would appeal to those trying to study aspects of physical matter to look within themselves as well as the subtler realms. We never will grasp the great design of nature from merely examining the physical shell. If the scientist will understand this basic idea, he will contribute greatly to the increase of real creativity in the scientific world.

Leichtman: You are almost suggesting that just as the proper way to examine a watch is to study the watchmaker, the proper way to study physical phenomena is to study the Creator and His designs and methods.

Einstein: I like that. Write it down.

Leichtman: And if physicists would view the phenomena of matter from this standpoint, they might learn more.

Einstein: Exactly. I hope I haven't spent too much time speculating on matter.

Japikse: Well, we will subtitle this interview, "What's the matter with matter?" *[Laughter.]*

Einstein: There you go.

Leichtman: Or, "Who's the matter with matter?"

Japikse: Or, "Why does matter matter?"

Einstein: Very good. It's been a pleasure being here and chatting about these ideas. I thank you for the opportunity.

Leichtman: And we thank you for coming.

Japikse: Yes, thank you.

GLOSSARY

ACUPUNCTURE: A healing technique developed by the Chinese, in which needles are inserted into specific points of a patient's body, for the purpose of relieving pain and curing diseases. The presence of the needles makes an impact on the etheric body, leading to the desired changes in the physical body.

ADEPT: One who is skilled in a specific talent. Esoterically, an adept has mastered the skills of soul consciousness.

ALCHEMY: The art of transmuting substances into more precious materials. As a science, alchemy was the precursor of modern chemistry. Esoterically, however, it is a practical system of magic and a symbol of the transmutation of the impurities of the personality under the guidance of the soul.

ANGEL: An entity belonging to the angelic or devic kingdom. Angels are not discarnate humans and have never been humans—they are part of a separate kingdom of life and have their own function and evolution. The angelic kingdom includes nature spirits, angels, and archangels.

ARCHETYPE: A basic pattern or ideal of creation. Archetypes are found at the abstract levels of the mental plane and are used by the soul as it creates the personality, its destiny, and its behavior.

ASTRAL MATTER: The substance or "stuff" of the astral plane.

ASTRAL PLANE: The plane of the emotions and desires. The astral plane is an inner world made of matter that is more subtle than physical substance, yet interpenetrates all physical substance. It is teeming with life of its own. The phenomena of the astral plane differ from physical phenomena in that they occur fourth dimensionally.

ASTRAL TRAVEL: Any form of movement of an entity on the astral plane. The uninformed sometimes use the term "soul travel" to describe this phenomenon, but this is incorrect: the soul pervades everything; it does not have to travel.

ASTROLOGY: The science of the interplay of cosmic energies. Astronomy is the science of the interrelationship of physical bodies and energies in the universe; astrology is the science of the interrelationship of *all* bodies and energies in the cosmos—astral, mental, and divine ones, as well as physical.

ATLANTEAN: A long stage in human development and civilization which ended catastrophically, through the misuse of its technological achievements. While the focus of growth for the average Atlantean was the nurturing and expression of the emotions, the science of that time reached heights not yet attained in our modern civilization. The period is called "Atlantean" because it was primarily centered on a now-submerged continent in the Atlantic Ocean and was referred to by this name by Plato. The heyday of Atlantean culture lasted from 100,000 to 12,000 B.C.

AURA: The light observed by clairvoyants around all life forms. It emanates from the surface and interior of the etheric, astral, and mental bodies. Clairvoyant observation of the aura can reveal the quality of health or consciousness.

AUTOMATIC WRITING: A form of writing in which the pen or typewriter is controlled by an entity or force other than the conscious mind of the writer. The controlling entity could be the subconscious of the writer, an elemental, a spirit, or the soul of the writer. Unless practiced by someone who fully understands its purposes, automatic writing can be dangerous.

BILOCATION: The phenomenon of appearing in two separate locations of the physical earth at the same time.

CLAIRAUDIENCE: The capacity to hear nonphysical sounds.

CLAIRVOYANCE: The capacity to see or know beyond the limits of the physical senses. There are many degrees of clairvoyance, allowing the clairvoyant to comprehend forces, beings, and objects of the inner worlds normally invisible to the average person.

CONSCIOUSNESS: The capacity to know or be aware.

CONTROL: A spirit guide for a medium.

DE LA WARR BOX: An experimental machine designed to detect and broadcast psychic energies—for example, for the purpose of healing.

DESTINY: The combined plans and commitments that have been made by the innermost spirit of an individual or a group for its future. A destiny would include plans for creative fulfillment, achieving enlightenment, and the events of life. It is not imposed from without—it is formulated by spirit.

DISCARNATE: A human being without a physical body, living on the inner planes.

DEVA: A Sanskrit term for "angel."

ELEMENTAL: A primitive life-form on the astral or etheric plane—generally, a nature sprite.

ELVES: A type of nature sprite or elemental, usually associated with wooded areas.

ENERGY: Conventionally, the capacity of a physical system for doing mechanical work. Occultly, however, energy is not considered to be a capacity originating in any physical system. Energy is an impulse of life which exists independently of physical matter. As it interacts with physical matter, it creates and animates and sustains physical forms. There are many different expressions of energy through form, and many different levels of quality and dimension among energies. The types of energy which have been observed and catalogued by science to date are only a tiny fraction of the energies of life, and science's comprehension of thge principles, behavior, and potentials of energy is as yet greatly limited.

ENLIGHTENMENT: Focused in the light of the soul.

ESOTERIC: An adjective which refers to knowledge of the inner worlds and inner life. In this book, it is used to refer to the knowledge of spirit—and to the body of teachings known as the Ancient Wisdom.

ETHERIC MATTER: The substance or "stuff" of the etheric plane.

ETHERIC PLANE: The most subtle realm of the physical plane. It is composed of the finest grades of physical matter, exceeding even the "fineness" of gases. In physics, the term "plasma" would be used for this grade of matter.

EVIL: Anything which retards the evolution of human consciousness. Contrary to public opinion, evil is not measured by likes and dislikes. Unpleasant events often help us evolve, and are therefore benevolent. The indulgences one person gives another, by contrast, may be quite pleasant, but distinctly harmful.

FAIRIES: A type of nature sprite or elemental. Fairies are usually invisible, but can be seen clairvoyantly. They help plant life grow.

FIFTH DIMENSION: A realm of existence in which there can be five different planes of movement from a single point, each of those planes being separated by ninety degrees or its abstract equivalent. All physical solids are part of this larger, fifth-dimensional context. The impact of the fifth dimension on third-dimensional solids can be seen as changes in the shape of a whole species or class of objects. In other words, if the growth registered in a single tree is fourth-dimensional in nature, the evolutionary growth of the species to which that tree belongs is fifth-dimensional. Archetypal ideas, from which whole species and classes of objects are produced, are part of the fifth dimension. The human soul also exists in the fifth dimension, as does heaven. Therefore, the true creative genius works in the fifth dimension. It is not, however, a common enviornment for the average human being.

FOURTH DIMENSION: A realm of existence in which there can be four different planes of movement from a single point, each of these planes being separated by ninety degrees. All physical solids are part of this larger, fourth-dimensional realm. The movement of a fourth-dimensional solid through the physical plane would be recognized as a change in the apparent three-dimensional shape of that object, such as in the growth of a tree. We act in fourth-dimensional ways every day—by associating relevant memories to current experience, by speculating about our future, and by perceiving underlying motives and attitudes of other people. Loosely speaking, the astral plane is the fourth dimension.

FREE WILL: The human individual's capacity to determine his own destiny within the context of universal order, purpose, and law. It is the soul, not the personality, which has the opportunity to use free will in its fullest sense.

GNOMES: A type of nature spirit or elemental. Gnomes are usually invisible, but can be seen clairvoyantly. They work with forces that aid in the development of the mineral kingdom.

GOD: The Creator of all that exists, visible and invisible; the life principle and creative intelligence underlying all life forms and phenomena.

HEAVEN: The state of consciousness of the spirit. Heaven is not a place populated by those who have died; it is accessible to incarnate and discarnate humans alike. It is a state of mind. In heaven are located the archetypal patterns of all creation, as well as the ideal qualities of human expression.

HIERONYMUS DEVICE: See De La Warr box.

HIGHER SELF: The animating principle in human consciousness—the inner being or soul. It is the guiding intelligence of the personality, the part of the human mind that is immortal.

HOROSCOPE: A chart depicting the interrelationship and relative strength of astrological forces at the time of birth for an individual, group, or nation. Used wisely, a horoscope can lead to a greater understanding of how cosmic forces influence creative manifestation.

HYPNOSIS: A psychological technique for communicating more directly (and sometimes more forcefully) with the subconscious of an individual. It is an artificial technique that does not make contact with the inner self of the individual.

I CHING: The Chinese Book of Changes, a system of pictograms and associated commentaries that are selected by the random fall of coins or sticks. It is a system of practical philosophy that lets the user tap an immense structure of thought to make sense of life.

INCARNATION: The period of time in which a human spirit is expressing itself through a personality.

INNER PLANES: A term used to refer to any one of several inner worlds or levels of existence, all of which interpenetrate the dense physical plane. Each physical human being exists on these inner planes as well as on the physical level, by dint of having bodies composed of matter drawn from them.

INNER SELF: The essence of the human consciousness which is the guiding intelligence of the personality. It is associated with the immortal aspect of the human mind.

INTUITION: In general usage, the capacity to know something without using the physical senses.

KARMA: A Sanskrit word meaning "reactiveness." Every one of our actions, thoughts, and emotions produces a reaction of like quality, sooner or later. Good deeds and thoughts produce beneficent reactions; cruel and selfish deeds and thoughts produce restrictive reactions. By dealing with these karmic effects, we gradually learn the lessons of maturity.

KIRLIAN PHOTOGRAPHY: A special type of photography which simulates etheric auras around objects.

MAGIC: In its original sense, the acts of a Magus or wise person with conscious awareness of the inner life of spirit. Pure magic, therefore, is the focusing of creative energies for the transformation of forms.

MATTER: The substance of life—energy in manifestation. There is mental and astral matter as well as physical.

MATERIALISM: The belief that the physical plane is the only plane of existence, or at least the most powerful and important. Materialism denies the importance of the soul, the existence of divine intelligence, and the invisible realms of life.

MEDITATION: An act of mental rapport in which the ideals, purposes, and intents of the inner life are discerned, interpreted, and applied by the personality. To be meaningful, meditation must be a very active state in which creative ideas, new realizations, and healing forces are discovered, harnessed, and applied to the needs of daily life. The current belief that meditation is a passive state of emptying the mind, by just sitting, is the antithesis of true meditation.

MEDIUM: A person who practices mediumship—the phenomenon of a nonphysical intelligence, usually a discarnate human, assuming some degree of control of a physical body in order to communicate something meaningful and useful.

MENTAL PLANE: The dimension of intellectual thought. One of the inner planes of existence, it also interpenetrates the dense physical plane. It teems with active life of its own, in addition to providing the substances for the mental bodies of all humanity.

METAPHYSICS: The philosophical and intellectual inquiry into the spiritual nature of all things.

MIND: The portion of the human personality that has the capacity to think. The mind is an organized field of energy which exists in invisible dimensions. It is *not* the physical brain, although it can operate through the brain.

MYSTIC: One who loves, reveres, and *finds* God and His entire Creation.

NINE UNKNOWN: A small group of physical people who, in effect, own the planet. They secretly oversee major events of civilization and protect occult knowledge. The label comes from a novel by Talbot Mundy, *The Nine Unknown.*

OCCULT: The hidden secrets of nature.

OUIJA BOARD: A silly device sometimes used by amateurs to contact "spirits"; it is neither a legitimate mantic device nor a safe alternative to mediumship.

OVERMIND: The one universal intelligence which pervades and inspires all individual minds. In other words, God. Also called the Oversoul.

PARAPSYCHOLOGY: The study of psychic phenomena.

PENDULUM: A device similar to a Ouija board—and equally dangerous.

PERSONALITY: The part of the human being that is used for manifestation in the earth plane. Composed of a mind, a set of emotions, and a physical body, it is the child of the soul and its experiences on earth.

PETER PAN DEPARTMENT: The realms of fantasy on the astral plane. Fantasies, wishes, and dreams that are generated here may appear real but have no substance.

PLANE: An octave in consciousness. All planes of consciousness interpenetrate the same space; they differ from one another in the quality of their substance. The human personality exists in the physical, astral, and mental planes.

POLTERGEIST: Literally, a "noisy ghost," so named because it makes itself known by loud noises and violent actions, such as propelling furniture across a room. In most cases, however, poltergeists are produced by subconscious projections of the emotions of physical people.

POSSESSION: A condition in which a human personality is controlled and used by an entity other than the soul that created it, against that soul's will.

PSYCHIC: A person who is able to perceive events and information without the use of the physical senses. The word is also used to refer to any event associated with the phenomena of parapsychology.

PSYCHOKINESIS: The phenomenon of influencing the movement of physical objects psychically.

RADIESTHESIA: The practice of diagnosing physical disease by psychically detecting radiations from affected organs and cells.

REINCARNATION: The process of the soul evolving through a successive and progressive series of different physical personalities.

SEANCE: The event of a discarnate communicating with physical people through a medium.

SITTING: A seance.

SOUL: The individualized principle of consciousness and creativity within the human being. It is the soul that evolves and acts; it is the soul that creates the potential of the personality, vivifies it, and guides it through certain life experiences designed to increase competence in living. The soul is a pure expression of love, wisdom, and courage.

SPIRIT: In this book, a word used primarily to describe the highest immortal, divine essence within the human being. Both incarnate and discarnate humans alike possess this spirit within them. In popular usage, however, the word is used to refer to the portion of the human being which survives death.

SPIRITUALISM: A religious movement that incorporates mediumship as a central feature of its worship.

SPOOK: An affectionate term for a discarnate.

SUBCONSCIOUS: The part of the personality that is not being consciously used at any given moment.

SUBPLANE: A subdivision of a plane. Occultly, there are seven subplanes in each plane of existence.

SUBTLE BODIES: The intangible "bodies" of thought and emotion all humans, plants, and animals have.

TELEPATHY: Direct mind-to-mind communication. Most often, telepathy occurs on the astral plane.

TELEKINESIS: The movement of a physical object through space by telepathic means.

THOUGHT-FORM: Literally, the form a thought takes on the plane on which it is created, usually the astral or mental. Visible only to clairvoyants, thought-forms are nonetheless created by every human being during the ordinary processes of thinking and feeling.

TRANCE: A state in which ordinary consciousness is

quieted so that another element of consciousness can use the physical voicebox and body.

UNCONSCIOUS: The part of the mind not ordinarily accessible to the conscious mind.

VAMPIRISM: The phenomenon of sapping or stealing energy from another individual through a telepathic or psychic contact. It is a common feature of sexual abnormalities, but can be found in other contexts as well.

FROM HEAVEN TO EARTH

The Hidden Side of Science is one of six books in the *From Heaven to Earth* series. Each book contains four interviews between Dr. Robert R. Leichtman and the spirits of prominent psychics, geniuses, and world leaders. They may be purchased individually for $11.95 apiece (plus $1.50 shipping) or as a complete set for $50, postpaid. The other five books in the set are:

The Psychic Perspective—Edgar Cayce, Eileen Garrett, Arthur Ford, and Stewart White.

The Inner Side of Life—C.W. Leadbeater, H.P. Blavatsky, Cheiro, and Carl Jung and Sigmund Freud.

The Priests of God—Albert Schweitzer, Paramahansa Yogananda, Andrew Carnegie, and Sir Winston Churchill.

The Dynamics of Creativity—William Shakespeare, Mark Twain, Rembrandt, and Richard Wagner.

The Destiny of America—Thomas Jefferson, Benjamin Franklin, Abraham Lincoln, and a joint interview with seven key spirits from American history—Alexander Hamilton, Franklin, Jefferson, the two Roosevelts, Harry Truman, and George Washington.

Orders can be placed by sending a check for the proper amount to Ariel Press, 14230 Phillips Circle, Alpharetta, GA 30201. Make checks payable to Ariel Press. Foreign checks should be payable in U.S. funds. In Georgia, please add 6% sales tax.

It is also possible to order by calling toll free 1-800-336-7769 between 8 a.m. and 6 p.m. Monday through Thursday and charging the order to VISA, MasterCard, or American Express.

OTHER BOOKS FROM ARIEL PRESS

Ariel Press is one of the leading publishers of fine books on the mind, psychic development, spiritual growth, healing, and creativity. Any of the following books can be ordered by writing directly to Ariel Press, 14230 Phillips Circle, Alpharetta, GA 30201, or calling toll free at 1-800-336-7769. Be sure to send a check in U.S. funds—or charge your order to Master Card, VISA, or American Express.

Add $1.50 per book for shipping (maximum shipping charge: $5).

ACTIVE MEDITATION
by Robert R. Leichtman, M.D. and Carl Japikse
$19.95

FORCES OF THE ZODIAC
by Robert R. Leichtman, M.D. and Carl Japikse
$21.50, hardbound

THE ART OF LIVING (5 VOLUMES)
by Robert R. Leichtman, M.D. and Carl Japikse
$40

THE LIFE OF SPIRIT (3 VOLUMES)
by Robert R. Leichtman, M.D. and Carl Japikse
$27

THE LIGHT WITHIN US
by Carl Japikse
$9.95

EXPLORING THE TAROT
by Carl Japikse
$10.95